ALLY VERSUS ALLY

ALLY VERSUS ALLY

America, Europe, and the Siberian Pipeline Crisis

Antony J. Blinken

PRAEGER

New York
Westport, Connecticut
London

Library of Congress Cataloging-in-Publication Data

Blinken, Antony J.
  Ally Versus Ally.

  Bibliography: p.
  Includes index.
  1. Gas industry—Political aspects—Soviet Union.
2. Gas industry—Political aspects—Europe. 3. Gas,
Natural—Soviet Union—Pipe lines. 4. Gas, Natural—
Europe—Pipe lines. 5. Soviet Union—Foreign economic
relations—Europe. 6. Europe—Foreign economic
relations—Soviet Union. 7. United States—Foreign
relations—Europe. 8. Europe—Foreign relations—
United States. I. Title.
HD9581.S652B58   1987   388.5   86-25222
ISBN 0-275-92410-6 (alk. paper)
ISBN 0-275-92616-8 (pbk)

Library of Congress Catalog Card Number: 86-25222
ISBN: 0-275-92410-6
ISBN: 0-275-92616-8 (pbk)

First published in 1987

Praeger Publishers, 1 Madison Avenue, New York, NY 10010
A division of Greenwood Press, Inc.

Printed in the United States of America

The paper used in this book complies with the Permanent Paper Standard issued by the National Information Standards Organization (Z39.48-1984).

10 9 8 7 6 5 4 3 2 1

Preface

At the threshold of its fifth decade, the Atlantic Alliance is showing serious cracks. On a number of seemingly unrelated fronts, the United States and Western Europe are at each other's throats. Bickering over government subsidies that distort agricultural trade, an unstable dollar, unfair competition for traditional export markets, and a mounting protectionist sentiment have pushed the allies to the brink of economic warfare. On military and strategic issues, similar tensions have surfaced. Taken totally by surprise, leaders across the Atlantic reacted with horror to the proposed limitation of nuclear weapons in Europe which the United States and the Soviet Union began to discuss at the Reykjavik summit. The so-called zero-option, originally proposed by Ronald Reagan and now taken up with a vengeance by Moscow, has sown confusion and concern on the Continent. The political Right, fearful of decoupling, sees the proposal as the first step toward another Munich, while the political Left, increasingly emboldened to knock NATO, is indulging pacifist aspirations and contemplating unilateral disarmament.

Faced with these problems, American policymakers seem tempted to throw up their hands and let Europe fend for itself. Republicans and Democrats alike are tired of seeing the U.S. devote almost one half of its defense budget to NATO and receive little more than complaints in return. Washington is also angered—although since the embarrassment of Irangate its criticisms are muted—by the Europeans' unwillingness to cooperate in the fight against terrorism. More generally, a new climate of isolationism is in the air—a belief that Europe is becoming less relevant, that American attention would be better devoted to the Pacific basin and Latin America.

In this context of heightened U.S.—European frustrations, the East-West trade issue is about to reemerge as the principal source of friction. Mikhail Gorbachev needs to revive commerce with the West if his ambitious plans to rebuild the Soviet economy are to be realized. The USSR has already made known its interest in joining world-wide economic institutions such as the International Monetary Fund and the General Agreements on Tariffs and Trade. In January, 1987, Moscow promulgated a law to promote joint ventures with foreign capitalist firms, allowing them for the first time to acquire equity holdings in Soviet industry. The Kremlin's new policy of *glasnost* and its gesticulations in the sphere of human rights, including the liberation of prominent dissidents, are also aimed, in part, at the expansion of commercial, scientific, and technological ties with the West.

Placed on the backburner after the pipeline crisis, the allied debate on trade with the East is starting to heat up again. West German foreign minister Hans-Dietrich Genscher's call for the West to help the Soviet Union modernize its economy was headline material throughout Europe. In the United States, concern about the costs to America's economy of trying to restrain trade with the USSR is also becoming part of the political debate. A widely-publicized report by the National Academy of Sciences concluded that efforts to keep high technology from the Soviet bloc have hindered American competitiveness, cost thousands of jobs, and resulted in billions of dollars in lost trade without significantly improving national security. This finding strikes home at a time when our foreign trade deficit has reached the unprecedented level of $170 billion and when we hold the dubious record of being the most indebted nation in the world.

The pages that follow focus on the fundamental discord between the United States and its western European partners on the central question of East-West trade. This source of incessant strain on the alliance culminated in the Siberian pipeline crisis of the early 1980s. Yet the onslaught of subsequent events has demonstrated that, far from being a closed episode, the pipeline affair was but the malignant symptom of disease that could irreparably harm the alliance. In order to survive both the recurring divisions among its members and the relentless challenges of its adversaries, the Atlantic community must face the experience and warning provided by the most dangerous rift in its history.

In today's volatile and interdependent international environment, the threats to the Atlantic partnership are complex and subtle. Direct military aggression is no longer the primary concern. More immediately disturbing is the USSR's ability to exploit contradictory Western interests and policies that have a potential for being as harmful to NATO as to the Warsaw Pact. Drawing on the pipeline conflict as a case study of the most persistently divisive alliance issue, this book hopes to show how the evolution of non-strategic East-West trade can proceed without constantly pitting ally against ally.

Acknowledgments

This book began at Harvard University. My adviser, Robert D. Putnam, and my readers, Richard N. Cooper and Charles Schilke provided the encouragement and guidance that enabled me to mold it into its present form.

I owe a great debt to my co-editors at the *Harvard Crimson*, who taught me the rudiments of writing as we put out Cambridge's only breakfast-table daily. Martin Peretz let me spend a year at *The New Republic,* whose brilliant staff continued my education as I struggled with the manuscript.

Several people have awakened my interest in foreign relations, offered me unusual opportunities for dialogue, and helped me gain access to some of the key players in the pipeline drama. I must single out Jean-Jacques Servan-Schreiber, Richard Gardner, the late Joseph Kraft, the late Jacob Javits, the late M. H. Blinken, Michael Thomas, and Daniel Patrick Moynihan in whose office I was privileged to spend an exciting summer.

A number of political leaders, government officials, businessmen, and scholars in the United States and in Europe who so patiently submitted to interviews and questions should be absolved of all blame for the content of this book. Among them are Helmut Schmidt, Richard Pipes, James Buckley, Wilhelm Christians, Marshall Goldman, Meyer Rashish, Robert Hormats, Michel Tatu, William Archey, Howard Lewis, Rudolf Augstein, Fabio Basagni, John Mroz, William Root, Michael Ely, Susan Haird, and Claus-Dieter Ehlermann. My gratitude also to those who agreed to be interviewed on the condition that they remain anonymous.

Most important, I thank my two sets of parents—Donald and Vera Blinken and Judith and Samuel Pisar. Their love, understanding, and wisdom constitute the foundation upon which this book was built.

Contents

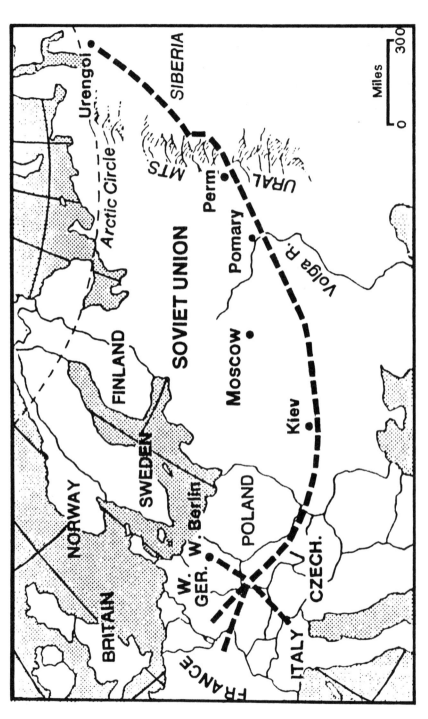

THE SIBERIAN PIPELINE PROJECT

STORM OVER
THE ATLANTIC

1

Pipe Dreams and Nightmares

"This day, June 18, 1982, could well go down as the beginning of the end of the Atlantic Alliance," France's foreign minister exclaimed before several dozen reporters crowded into an ornate room on the Quai d'Orsay. "The United States has just declared what amounts to economic warfare on her allies in western Europe." A chorus of similar indignation echoed in Bonn, London, and Rome. In Moscow and other capitals of the Warsaw Pact nations, the Communist leadership delighted in the furious diplomatic brawl that divided its adversaries.

Europe's outrage was triggered by a decision that had just been announced in Washington. President Ronald Reagan had prohibited the delivery of any American equipment and technology for the construction of a 3,000-mile pipeline to transport natural gas from Siberia to western Europe. Since most European firms involved in the project used components or know-how of American origin, their supply contracts were placed in jeopardy. Washington had left nothing to chance: even foreign-based subsidiaries of American companies using no American goods or technology whatsoever were enmeshed in the net of the embargo.

The largest East-West commercial venture ever undertaken, the pipeline project had been hailed in Europe as "the deal of the century." The Soviet Union was to supply five member countries of the European Economic Community (EEC) with close to one-third of their gas needs (about 6 percent of their overall energy requirements). Soviet gas would help the Europeans alleviate their dangerous and costly dependence on Middle East oil and the caprices of the Organization of Petroleum Exporting Countries (OPEC) cartel. European

3

companies stood to earn $7 to $10 billion from equipment sales. This would guarantee tens of thousands of jobs and keep a number of Europe's largest industrial manufacturers afloat. The prospect of job creation inherent in the multibillion-dollar equipment orders was an irresistible temptation for countries whose industries were mired in recession and suffering unemployment levels unmatched since the Great Depression.

With such high stakes in play, the violent European reaction to the American embargo was not entirely surprising. The indignation reflected, however, more than just legitimate concern over energy diversification, jobs, and profits. Washington was, in effect, forcing sovereign allied states and independent foreign companies to do what they perceived to be against their own national and business interests. The embargo was an intolerable affront to the Europeans, for, in their view, it brazenly asserted the United States' right to make trade and foreign policy for its allies—whether they liked it or not.

Since the October Revolution the Soviet Union's attitude toward trade with the capitalist world has gone through many contradictory stages. During Lenin's New Economic Policy, the USSR enjoyed a brief but intense flirtation with Western industrialists. Stalin eventually turned inward in an effort to achieve economic autarky, at first within the confines of the USSR and then, after World War II, within a Communist bloc enclosing the newly subjugated countries of eastern Europe. But under the impact of insoluble domestic problems, isolationism gradually gave way to increasing involvement in the world market. An agricultural system in permanent crisis has forced Moscow to purchase vast quantities of grain from the West every year. A generally inefficient industrial system, at least in the nonmilitary sector, has made it necessary to import advanced equipment and technology as well.

The Soviet Union is blessed with enormous resources, both human and material. Yet it has been unable to put these resources to efficient use. It is as though Communism, when applied to economic endeavor, succeeds only in crushing the creative spirit of the people it touches. The USSR has shown no aptitude for producing finished products capable of competing in the sophisticated markets of western Europe, North America, and Japan. As a result, it is reduced to exporting its raw materials in order to earn the hard currency needed to finance its extensive shopping list. Gas, oil, ore, gold, diamonds, and timber—these are the principal assets of the Soviet state trading enterprises—assets not produced by the genius of Socialists but taken from the soil of Mother Russia. They have been virtually exhausted in the European part of the country. The Kremlin has therefore turned its attention to the immense riches buried in the great wastelands of the northeast.

Siberia is a continent-sized landmass floating on a sea of natural gas and oil. But drawing resources from its inhospitable soil is no simple task. The average temperature in the region lies well below zero centigrade, making for dismal work conditions and requiring costly investments. Much of the land stays frozen most of the year; when the thaw arrives, it creates rivers of mud that carry away the structures and equipment used in the extraction of raw materials. The problem is compounded by the Soviets' inability to manufacture state of the art oil and gas production equipment. The rudimentary technology needed to overcome the Siberian elements exists in the USSR, but it is better and more cheaply produced in the West.

From the Soviet point of view, western Europe makes an ideal market for Soviet energy exports. With some significant exceptions, European nations have little indigenous energy production. In the early 1960s, a classic pattern of economic complementarity began to emerge between the two segments of the continent. The west Europeans would trade technology and equipment to the Soviets, provided on generous credit terms, in return for Siberian oil and natural gas. East-West energy trade quickly became an important factor in the world economy. Within ten years, the USSR was a major exporter of oil and gas, thanks to a production and transportation infrastructure built with substantial assistance from the West.

In the early 1980s, Soviet planners decided to devote considerable effort to expanding natural gas production in Siberia. Moscow was especially interested in increasing gas exports to the West in order to make up for an anticipated drop in oil production. Western technology would be needed if Moscow was to develop some of the more inaccessible but energy-rich areas of Siberia. The Europeans, traumatized by the oil shock of 1973 and struck once more by an OPEC-induced jump in the price of oil in 1979, were glad to buy more Soviet gas and thus diversify their energy sources. Both sides welcomed the idea of dramatically increasing the USSR's capacity to export gas from Siberia to western Europe.

Although of unprecedented scale, the Euro-Siberian deal fit what had become a model for East-West commercial intercourse. Apart from the technology-for-raw-materials equation, such trade depends on the West's willingness to engage in barter and counterpurchase transactions, and to grant the Soviets credit at low interest rates. Strapped for hard currency, the USSR would have great difficulty financing its purchases of equipment by other means. The Europeans believe that generous credit terms employed to promote exports to any buyer are in their economic interest. Without such credit, there would be no foreign equipment orders of great magnitude. For this reason, Washington's criticism of the pipeline's generous financing package was perceived in the European capitals not only as an attack against a particular deal, but also as a blow to the very logic of East-West commerce.

The Reagan administration had a decidedly different view of the Euro-Siberian project. Officials in the Department of Defense and on the National Security Council (NSC) staff feared that western Europe was subjecting itself to potential energy dependence and to dangerous political leverage by relying on the USSR to supply so much of its gas. The Americans claimed that some of the technology furnished for the pipeline would be of strategic-military value to the USSR. They criticized the economics of the deal, insisting that the Europeans had been outnegotiated by Moscow. And they decried the alleged use of forced labor for the pipeline's construction.

Equally important, Washington saw in the pipeline an opportunity to put a new American strategy for relations with the Soviet bloc to the test. The Reagan team had come to power in 1981 convinced that the policy of détente instituted by President Nixon was a failure. Not only had Moscow continued to intervene throughout the world as it saw fit, but it had built its military machine to a point of near parity with the United States.

But there was a flip side to Soviet adventurism and the growth of the Red Army of which Washington was well aware. The USSR was encountering increasingly pronounced economic difficulties that seemed endemic to its system, and not simply the result of a cyclical recession. The Reagan administration reasoned that the West should do nothing to help the USSR out of its dire economic straits. Instead, why not let the Soviets bear the full burden of their systemic failure? At the very least, this might force Moscow to reduce military expenditures. In an effort to avoid collapse, it might even require the USSR to dramatically restructure its economy—with all the implications such a change could have for the political and social organization of the Soviet state and the long-term ideological contest between communism and capitalism. For a number of influential policy makers, the time seemed opportune to engage the Soviet Union in a round of economic warfare, regardless of the price that might have to be paid by the economies of western Europe and the American business community, which would be excluded from Soviet bloc markets.

From Washington's perspective, the pipeline project was a glaring example of just the sort of "help" that the West should avoid giving to the Soviets. The deal included several billion dollars of credit at below-market interest rates to enable the USSR to purchase Western equipment. Moscow stood to earn substantial sums of hard currency from the sale of its gas. Washington assumed that this would make it easier for the Soviets to divert to the military resources that otherwise would have been used to prop up the economy or to purchase Western technology that Moscow needed and could produce itself only with great effort. The expansion of the USSR's energy infrastructure would also be a long-term boon to the Soviet domestic economy. In short, the pipeline should not be built because it was incompatible with the policy of trade denial that Washington sought to implement against the Soviet Union.

This form of economic warfare was not a novel strategy. In varying degrees, it characterized the American approach to the Soviet Union from the end of World War II until the late 1960s. The policy owed much to America's lack of natural incentives for trade with the USSR. Washington tended to see commerce with the East as a zero-sum game: since it helped them, it by definition hurt us. Trade with the USSR was viewed by many as giving aid to the enemy.

For decades, Washington was able to set the Western nations' trade policy toward the East. Yet America's position in the world has changed markedly since the introduction of export restrictions, import discriminations, and credit limitations against the Communist countries in the late 1940s and early 1950s. The United States no longer boasts a quasi monopoly on the development and production of sophisticated industrial equipment and technology. For its restrictive trade policies to succeed, Washington requires cooperation from other nations, especially its industrialized Western allies.

Here, too, the evolution of geopolitics and the world economy has run counter to American interests. Within the alliance, the competitive edge in many industries has shifted in favor of Europe and Japan. These countries had been dependent on the United States for protection and economic aid during the postwar decades, and were thus quick to acquiesce to Washington's views on East-West policy. Once the allies had rebuilt their economies, they no longer felt compelled to play follow the leader in Washington. As early as the 1960s, the western Europeans began to nurture close economic ties with the Soviet Union and its satellites. The United States' brief espousal of détente and increased trade with the East left the allies free to expand their own commerce eastward. By 1982, while American trade with the USSR totaled just $2.5 billion, western European trade with the Soviet Union stood at $41 billion.

The Europeans' view on trade relations with the USSR owes much to their geographic proximity to that country and their role as a natural economic complement to the East. America's allies have always felt a strong need to coexist with the Soviets. At the same time, as trading states that, unlike the United States, devote a relatively large share of their GNP to international trade, the Europeans have been reluctant to sacrifice the benefits of commerce for what they see as dubious strategic propositions. The West, they insist, cannot strangle the Soviet Union economically. Open trade, restrained only by good business sense and strategic prudence with regard to the exportation of militarily useful goods and technology, should be the basis of the East-West commercial relationship.

Among those who favor expanded East-West contacts, some believe that commerce will eventually have a moderating influence on Soviet domestic and foreign policies. Like the economic cold warriors, they too would like to see a change in Soviet behavior, especially in the domain of human rights at home and diplomatic or military opportunism abroad. While the ends are similar, the means could not be more different. Those who support trade want to integrate the Soviet Union into the world economy. In their opinion, the USSR's economic shortcomings and the ongoing generational change in the Kremlin leadership present new opportunities for constructive coexistence and peaceful competition. A rising class of young technocrats in the Mikhail Gorbachev mold, scornful of the stagnation imposed by the Stalinist model, will seek out trade and broadened industrial intercourse with the West to improve the Soviet economy. From the West's perspective, the growth of economic exchange would, it was hoped, stimulate positive change in the Soviet political and social systems, and restraint in its international behavior.

This divergence of policies has been simmering within the alliance for years. The two antagonistic trade philosophies—the "positive" one more prevalent in Europe, and the "negative" one more prevalent in the United States—have given rise to a number of minor crises from time to time. In 1962, for example, a dispute strikingly similar to the Euro-Siberian affair arose when Washington forced West Germany to cancel the sale of steel pipe to the USSR for the construction of a pipeline to supply Moscow's eastern European satellites with oil. However, the Nixon/Kissinger détente instituted ten years later changed—albeit temporarily—the course of events. Washington envisaged twin pillars of support for East-West coexistence: arms reductions and political dialogue on the one side, and expanded economic ties on the other. Kissinger, the policy's prime architect, saw the economic relationship in an inherently geopolitical fashion. Commercial relations were to be employed as either a carrot or a stick—that is, liberalized when the Soviets behaved, restricted when they did not. Because this required a meaningful volume of trade between East and West, the new American approach was aligned with European thinking.

The Siberian pipeline grew out of détente. Ironically, its roots go back to an abortive project in the early 1970s that would have brought large quantities of Soviet gas to the shores of the United States itself. As the pipeline contracts were in the process of negotiation at the outset of the 1980s, though, détente fell from grace and the Reagan administration seemed determined to return to a form of economic warfare. The pipeline was too huge a project, too symbolic of the era of détente, and too contrary to the principles of trade denial to escape American opposition. Faced with western Europe's determination to proceed with the venture, Washington could not simply order that the contracts be canceled. A cogent reason to take action was needed, and history saw fit to provide it.

Two weeks before Christmas 1981, Gen. Wojciech Jaruzelski, the Polish prime minister, declared martial law and established a "military government of national salvation." He had the leaders of the Solidarity trade union arrested in order to subdue a spreading movement of popular dissidence supported by the Catholic Church. Government officials in the Western capitals, although not surprised, were unsure how to respond. While it was apparent that Moscow was intimately involved in the crackdown, Soviet troops had not entered Poland. What was to be done to express the West's condemnation of these repressive initiatives?

Over the following days, it became clear that the United States and its European allies had very different ideas concerning the best policy to adopt in response to the crisis. Washington urged a coordinated set of economic sanctions aimed at Poland and the Soviet Union. The Europeans believed that sanctions would not work. They argued, moreover, that martial law, however distasteful, was better than a Red Army invasion and, in any event, an internal matter for the Poles to resolve by themselves.

President Reagan, frustrated by his inability to command allied cooperation, decided to take unilateral action. On December 29, he announced a series of sanctions against the Soviet Union. The list included a number of predictable measures, such as suspension of Aeroflot's landing rights within the United States, the postponement of negotiations on a maritime accord and a new long-term agreement to sell grain to Moscow, the closing of the Soviet Purchasing Commission office in the United States, and an indefinite halt to the issuance of licenses for the export of high technology goods.

It was, however, another sanction that attracted the most attention. Reagan ordered that special government licenses be required to sell the Soviets an expanded list of equipment for oil and natural gas production. Furthermore, he simultaneously suspended any such licenses already granted. American manufacturers of equipment used for oil and gas extraction, including Caterpillar and General Electric—the president's former employer—quickly complained that the embargo would cost them several hundred million dollars in lost export contracts. In London, Paris, Bonn, and Milan, western European companies and foreign ministries also made known their alarm. What effect, they demanded to know, was the embargo expected to have on the Siberian pipeline project?

The European's anxiety and confusion were understandable. Washington had failed to make clear whether the embargo applied to European pipeline contractors incorporating American technology into their finished products and to those producing equipment under U.S. license. Nor could anyone say with certainty whether existing contracts were included or only those to be concluded in the future. If the embargo proved to affect European companies and preexisting contracts, the Europeans would be in the uncomfortable position of either going ahead with the pipeline in defiance of Washington's directive or docilely obeying the White House and thus abandoning the project.

The Reagan administration itself was divided over the scope of its embargo. The difference in views was based primarily on disparate perceptions of the goals to be achieved: while the State Department understood the embargo to be simply a response to the Polish crisis, and thus limited to American companies and future contracts, Defense Department and NSC staff strategists, whose aim was to stop the pipeline, sought to make the embargo as broad as possible.

In the absence of a clear and definitive statement of U.S. policy, the extent of the embargo remained uncertain throughout the winter and spring of 1982. While the Europeans pressed ahead with the pipeline, Washington attempted to use this uncertainty to obtain other concessions from its allies in more general areas of East-West trade policy. The administration was especially concerned by the ease with which Moscow obtained extensive credits at low interest rates from European bankers and government export subsidy programs. Washington therefore threatened to extend the pipeline embargo if these credits were not cut back. As the Versailles summit of the West's most industrialized nations approached in early June, an agreement within the alliance calling for less credit to the East in exchange for an American "hands-off" policy on the pipeline had seemingly been reached. But the agreement fell apart almost as soon as the summit ended. On June 18, 1982, President Reagan made his decision to broaden the embargo.

The crisis within the alliance reached its peak in August when the French, German, British, and Italian governments all ordered pipeline suppliers in their respective countries to disregard the American embargo and fulfill their contractual obligations to the Soviet Union. As ships sailed from half a dozen European ports carrying the embargoed equipment to the USSR, Washington responded by denying the defiant western European companies access to all American goods, services, and technology, and threatening criminal proceedings.

By November, it was evident that the embargo was ineffective. It not only had failed to prevent the Soviets from acquiring most of the needed Western equipment but also had dangerously strained the alliance. Tensions reached such a point that Washington feared the pipeline fiasco would weaken European support for NATO's deployment of cruise and Pershing II missiles in answer to the SS-20s that the Soviets had aimed at western Europe. While the Europeans saw themselves as the victims of a "trade war" in which the United States was the aggressor, a number of influential American legislators complained about the failure of the NATO allies to assume an adequate share of the common defense burden and hinted at a withdrawal of American troops. The situation was rapidly becoming embarrassing and unmanageable. Six months after he extended the embargo, Reagan reversed himself and had it rescinded.

In early 1984, the Soviets triumphantly announced that the pipeline had been completed more or less on schedule. Today, gas flows unimpeded through the Euro-Siberian, heating homes and powering industries all over western Europe. The acrimonious alliance dispute seems to have faded from the West's collective consciousness. But in fact, the divisive issues that paralyzed the alliance during the summer of 1982 were simply papered over and remain unresolved. Unless these issues are thoroughly addressed, it is only a matter of time before a new crisis shakes the alliance.

This is a predicament the Atlantic community must strive to avoid. George Kennan wrote 40 years ago that the East-West competition is a "long-range fencing match in which the weapons are not only the development of military power but the loyalties and convictions of hundreds of millions of people and the control or influence over their forms of political organization. . . . It may be the strength and health of our respective systems which is decisive and which will determine the issue."[1] Four years after the pipeline crisis, Kennan's words remain applicable to the Atlantic alliance.

In the wake of the Reagan-Gorbachev summit, this is an especially inopportune moment for the West to be divided on commercial policy toward the East. Trade is once more coming to the foreground as a central issue in East-West relations. If the ambitious economic program announced by Mikhail Gorbachev at the 27th Communist Party Congress is to be fulfilled, the Soviets will again tempt the western European, Japanese, and American business communities with substantial export opportunities. Should the West accept this offer? The dispute over the pipeline highlights many of the issues raised by this deceptively simple, but fundamental, question.

With the help of hindsight, it is today easier to determine whether the construction of the Euro-Siberian pipeline was a wise or a foolish strategic and economic undertaking for the West. This aspect of the inquiry poses a number of questions, starting with the need for the pipeline, the extent of European dependence on Soviet gas, the size and significance of Moscow's expected hard currency earnings, the implications of helping the USSR exploit its natural resources, the financial packaging of the deal, and the price of Soviet gas.

The second broad inquiry involves the long-term consequences of the crisis within the alliance. In 1986, four years later, its impact on relations between the United States and western Europe and on their respective approaches to East-West trade may be analyzed more rationally than in the heat of the moment. Some believe the crisis prompted the Europeans to rethink their ties to the East and promoted greater Atlantic cooperation in trade and nontrade areas. Others fear the embargo did lasting harm to American industry and its reputation as a reliable world-wide supplier of goods and equipment.

Finally, the pipeline affair offers new insights into the future possibilities and limitations of Western trade policy toward the Soviet Union. It is a fresh point of entry into the unresolved debate that has plagued the NATO alliance since its creation. What role does Western technology really play in enhancing Soviet economic and military power? How effective are national and multinational trade restrictions applied to the Soviet Union? What impact do such policies have on Western economies? Is a more coherent and coordinated Western strategy for trade with the USSR and its satellites possible?

A case study of the pipeline cannot help straying into the broader issues of East-West coexistence. Trade, it is true, constitutes only one piece of a puzzle that includes ideological conflict, territorial expansion, arms control, and human rights. Yet the division and confusion surrounding the pipeline affair mirror the ambiguous and contradictory Western responses to the Soviet challenge. The alliance has vacillated among trying to destroy the Soviet system by force, merely containing its expansion, and encouraging it toward peaceful evolution; the pipeline crisis is a useful litmus test for these approaches. Moreover, it highlights the practical limitations on each approach within the framework of a voluntary alliance composed of politically independent and economically competitive powers.

As a chapter in the life of the alliance, the pipeline saga makes fascinating history. It is living history that does not as yet permit final conclusions about the project itself or its economic and political consequences. The flow of gas has not reached full capacity. Prices and terms of delivery have been renegotiated and are scheduled for another major review before the end of the decade. The world energy market is highly volatile—as the West knows all too well. While the dramatic plunge in the price of oil and the projected development of Norway's immense Troll deposits now make Soviet gas appear far less important to western Europe than it did in 1980, there is no guarantee that the oil market will remain soft. Only the course of future events will fully demonstrate whether the great expectations, deep concerns and extreme tensions generated by the pipeline project were justified.

THE DEAL
OF THE CENTURY

2

Genesis of a Pipeline

Wilhelm Christians, president of the Deutsche Bank, has dealt with the Soviets since the early 1970s, when he arranged the financing package that enabled the USSR to purchase large quantities of steel pipe and other pipeline-related technology from German manufacturers that was used to produce and export gas to the West. As the head of the Federal Republic's largest bank, and as a member of the board of several of its most powerful companies—including the steel manufacturer Mannesmann—Christians can easily claim the title of Germany's premier financier.

Sitting in a sparsely decorated but imposing office overlooking the Koeningsallee in Düsseldorf, Christians explained why the Euro-Siberian project was so crucial to his country.

> Jobs were a significant factor, without a doubt. But more important was our great concern in Germany about energy supply. We had been seeking diversification of sources since the 1973 OPEC crisis after the Yom Kippur war. That was the motivating force that made building this pipeline so imperative.[1]

THE OIL PREDICAMENT

West Germany's energy predicament is typical of that of most European Economic Community (EEC) countries. In 1981, when negotiations regarding the pipeline were in their final stages, the Federal Republic consumed 262.9 million tons oil equivalent (mtoe) of primary energy, but produced only 130.5 mtoe. In other words, the

Germans had to import 50 percent of their energy requirements. Oil was a particularly dramatic problem. That resource satisfied half of West Germany's energy needs, yet more than 90 percent of the Federal Republic's oil was imported, mostly from OPEC. Similarly, France imported about 70 percent of its oil and Italy 80 percent; both countries depended on oil for well above half of their total energy consumption.[2]

The potential danger of such external dependence had been brought into sharp focus nearly a decade earlier. During the Arab-Israeli conflict which broke out in October 1973, OPEC used its oil leverage on the Western industrialized nations as a political weapon and a punitive measure against them for their support of Israel. The tide had dramatically turned from the days when the oil producing countries were seen by many as nothing more than mythical Araby. The Arab world, realizing the extent of its collective power, had banded together and turned the tables on the oil-consuming countries.

The United States and the Netherlands were the victims of a total OPEC embargo for their direct support of Israel, while West Germany, Italy and Japan suffered phased reductions of 5 percent per month. Only France and Great Britain were exempt from supply cuts.[3] After six months, OPEC lifted its embargo. But by then, the price of oil had escalated dramatically, from $2.70 per barrel in October 1973 to $10.46 per barrel in March 1974.[4]

The embargo underlined two fundamental aspects of energy dependence. The West had found itself at the mercy of (a) a particular resource: oil, and (b) an organized group of nations with extensive control over that resource: OPEC. OPEC supplied the EEC with more than 50 percent of its total energy needs, and Japan with more than 80 percent.[5]

The 1973 crisis proved to be only the first of several shocks the West would endure during the 1970s, all of which revealed its glaring inability to assure a free flow of oil at a reasonable price from the Middle East. The consequences were particularly alarming for the Europeans. Ever since the United States had halted Britain, France, and Israel's attempt in 1956 to reverse Nasser's seizure of the Suez Canal, European policy makers believed that Washington had implicitly assumed responsibility for the security of Western interests in the Middle East, including the energy lifeline.[6] The limits of America's ability or willingness to discharge this responsibility became obvious as the 1979 oil crisis followed the 1973 shock, pushing prices from $13 to $24 per barrel. That same year, the overthrow of the shah in Iran further disrupted supply. In both instances, the United States seemed incapable of stemming the turmoil in the Middle East.

Washington's failure to act resulted from a lack of "power in place" in the Middle East and, with the exception of Israel, from a dearth of stable allies in that part of the world. Even in the wake of the Soviet invasion of Afghanistan in 1979, when President Carter announced that any attempt by an outside force to gain control over the Persian Gulf would be regarded as an attack against the vital interests of the United States, subject to an American military response, America's allies remained unconvinced. Their skepticism proved to be well founded when an attempt to rescue American hostages in Iran in April 1980 ended in tragic failure.[7] "It became obvious to all of us in Europe that no one can guarantee the supply of Middle East oil," said former West German Chancellor Helmut Schmidt. "There are too many factors beyond our control, from a Kaddafi, to a Khomeini, to war between producer states like Iran and Iraq."[8]

As the West confronted its dependence on OPEC, a major drive to devise methods for conserving energy began.[9] In the aftermath of the first oil crisis, and at Washington's insistence, the Western industrialized nations and Japan formed the International Energy Agency (IEA) to coordinate supply policies. Certain nations, notably France, refused to join, partly for fear of endangering their friendly relations with the OPEC countries. All the consumer nations realized, however, that an "away from oil" strategy was of vital necessity. Conservation efforts, coupled with the start of a recession in the late 1970s, helped: oil consumption dropped by 3.6 percent in the EEC between 1973 and 1978. But by the early 1980s, the measurable results were mixed at best. Oil still accounted for half of the total energy consumption of the EEC, and the Community remained the largest oil importer in the world.[10] Analysts feared that the price of oil would rise well above $35 per barrel. Only by shifting consumption to other sources of energy could the Europeans hope to alleviate their oil supply dilemma.

THE NATURAL GAS ALTERNATIVE

Between 1974 and 1979, natural gas consumption in western Europe rose at an average annual rate of 4.7 percent, though total energy consumption increased by just 2.2 percent a year.[11] By 1981, natural gas constituted 17, 12, and 17 percent, respectively, of West German, French, and Italian energy consumption.[12] These figures attest to the growing importance of natural gas to western Europe at the start of the decade. The oil crises are the primary reasons for the EEC's increasing reliance on gas. But beyond that, gas possesses inherent qualities that make it an attractive source of energy.

Gas can replace oil in many areas, and can serve a wide variety of uses. It is employed as a raw material in the petrochemical and chemical industries, and has myriad industrial applications. Gas is a source of electricity, heating, cooling, and cooking. Its exceptional cleanliness, both relatively and absolutely, makes gas environmentally benign. Finally, it is a resource still in abundant supply and reasonably priced.[13]

Even in the absence of an "away from oil" strategy in the West, it seems likely that an evolution toward gas would have occurred in view of the problems involved in developing and producing other sources of energy. Coal is notoriously "dirty." Solar energy, in the minds of most people and in the budgets of most governments, lies in the distant future. Nuclear power, despite its tremendous potential, is the subject of impassioned controversy, and presents a political risk to those who seek its expansion. Events such as the Three Mile Island accident and the Chernobyl disaster, coupled with the nuclear arms race, have done much to fuel the fervent opposition of environmental and pacifist groups throughout the world.

The three major industrial regions of the Western alliance—the United States, Japan, and western Europe—have been forced to increase their natural gas imports in order to meet growing domestic demands. In the United States, the domestic production of gas is on the decline, despite partial deregulation of price controls. If, however, the Alaskan fields are more fully exploited, and price restraints are removed from gas found in "old" fields, the United States should remain largely self-sufficient. Japan, a country that is particularly poor in energy resources, has practically zero production. Western European output, with the exception of Norway and perhaps the Netherlands, will peak toward the end of the 1980s.[14] Although Great Britain, the Netherlands, and Norway produce enough gas for their own needs, Germany imports 67 percent of its gas, France 70 percent, Italy 55 percent, and Austria 64 percent. Of this, between one-third and one-half originates in sources outside the Organization for Economic Cooperation and Development (OECD) member states.[15]

Although most EEC members import significant amounts of gas from outside the Community, the risk of extensive "supplier control"—or, in other words, the potential for a gas cartel similar to OPEC—is limited. The natural gas market does not have the same characteristics as the oil market. To begin with, natural gas suppliers are geographically dispersed and politically disparate; they include Australia, Mexico, Malaysia, Iran, the USSR, Nigeria, Qatar, Algeria, Canada, the Netherlands, Great Britain, and Norway. The Royal Institute for International Affairs in London has concluded that

> There is no indication of any common ground for the formation of an exporters' cartel. The major actors are very different political entities, preoccupied with different aspects of the trade and with different aspirations.[16]

Despite the overall low cost of gas compared with that of oil, its production requires enormous initial investments. A natural gas liquefaction plant with an annual capacity of 10 billion cubic meters (bcm) currently costs about $1.5 billion, a tanker with 130,000 cubic meter (cm) capacity $150 million, and a gas processing plant with a 10 bcm/year capacity about $400 million.[17] A thousand-mile pipeline for "dry" gas requires several billion dollars' worth of steel pipe, compressor stations, and other processing equipment. These enormous capital commitments necessitate intense cooperation between suppliers and purchasers in the gas industry, with most contracts being drawn up to ensure shared risk taking.

As gas becomes more important, the suppliers' position inevitably will be strengthened. But so far, the inability of the more aggressive major producers, such as Algeria, to set a uniform international price for gas based on crude oil parity, and the continued abundance of gas, indicate that the threat of a suppliers' market lies far in the future.

For reasons of physical proximity, the EEC has relied mostly on gas supplies from its member states, African and Asian countries, and the Soviet Union. On other continents, Canada, Australia, and Mexico have substantial reserves. Technological progess in the means of production and transportation will eventually make this gas more accessible to the EEC.

Given the choice, the western Europeans would of course prefer to buy gas within the EEC, but the limits on Dutch, British, and, until 1986, Norwegian export capacity have made this policy unfeasible.

Iranian, Nigerian, Algerian, and above all Soviet gas has thus become increasingly important to the EEC.

WESTERN EUROPE'S SUPPLIERS

The Netherlands is the world's third largest gas producer after the United States and the Soviet Union, and has played an important role in supplying the other EEC members with gas. In 1976, exports to the EEC peaked at 53.4 bcm. By 1982, the Netherlands was the single largest supplier of gas to France and West Germany.

Until the 1984 discovery of additional deposits, however, the long-term production outlook was grim. The Dutch had calculated that, at 1982 rates of production, the theoretical life expectancy of their reserves was a meager 18 years. By comparison, Algerian and Soviet "life expectancies" are 126 and 75 years, respectively.[18] In an attempt to gain control over the situation and protect its national interests, the Dutch government embarked on a conservation policy at the start of the decade that included the nonrenewal of the major export contracts expiring in 2000.

It is now likely that new export contracts will be granted as a result of the additional gas discoveries.[19] The Europeans contracted for more Soviet gas via the Euro-Siberian in 1981/82, believing the Netherlands could no longer supply them on a long-term basis. Still, the amount of Dutch gas available for export during the 1990s is unclear, as the new discoveries may well be viewed by the Dutch as a reprieve for future domestic needs rather than as a license to continue export operations at peak levels.

Great Britain has also estimated the life expectancy of its gas reserves to be 18 years but, unlike the Netherlands, has made no new discoveries. According to the most recent IEA report, government policy concerning gas exports is in flux. Although the Thatcher government has promoted increased production, the prevailing strategy is to encourage imports in order to defer the depletion of domestic reserves. Production is thus a function not only of demand but also of the political relationship between export policies and domestic conservation needs.[20]

The fast approaching necessity of protecting domestic interests in the face of diminishing domestic reserves may soon simplify the equation. Maximum British production in 2000 should fall somewhere between 36 and 48 bcm/year.[21] This level will rule out substantial exports to the rest of the EEC.

Norway possesses large reserves with a life expectancy of 34 years. In the spring of 1986, the Norwegians decided to develop the Troll field, the largest source of gas in the North Sea, from which they are to supply France, Germany, Belgium, and Holland with 20 bcm per year by 2000. This should give Norway a 25 percent share of the EEC market, roughly equal to that of the Soviet Union.

The Troll project represents a surprising and abrupt shift in export policy. Back in the early 1980s, as the Siberian pipeline was being built, Oslo had adopted a go-slow approach to production in order not to disrupt economic development by overemphasizing the gas sector. Norway also hoped to preserve its gas for future use. Despite proven reserves of 1,400 bcm and total recoverable reserves of 4,114 to 5,318 bcm, Norway projected in 1980 that overall production would only reach 36 to 63 bcm by 2000.[22]

While the political winds have shifted, technological difficulties remain that could hinder the new North Sea development. The Troll field is located under 336 meters of sea water, which is more than twice the depth of the deepest field now being exploited. The sea bottom itself is unstable, as an accident on the Phillips Petroleum Ekofisk platform demonstrated in 1985. Phillips was forced to warn consumers that it might have to diminish oil and gas production in the North Sea.[23] The IEA's assessment of the situation stated that:

> North Sea gas development, due to the hostile physical environment, requires constant development of technology. This, combined with often complex reservoir configuration, may delay the achievement of maximum gas production levels and will result in high development costs.

Even if the Norwegians overcome their technological difficulties, the price of their gas is almost certain to be high compared to other sources. One reason Norway hesitated so long to exploit its North Sea reserves was because low gas prices made such exploitation uneconomic. Buyer countries in the Troll deal must have guaranteed Norway substantial remuneration in order to convince Statoil, the state-owned oil and gas company, to undertake the project. Indeed, a Statoil official says that the contract terms "are in line with the best prices we have seen."[24]

In Nigeria, important reserves were discovered during the 1970s, and now total about 1,455 bcm.[25] But a major liquefied natural gas

(LNG) export project was deferred by the government in 1981 because the high cost involved—between $10 and $14 billion—was considered prohibitive. In addition, the government removed a large allocation of funds for gas production and export from the Fourth National Development Plan (1981–85). For these reasons, and because Western developers are hesitant to invest in this politically unstable country, export potential remains uncertain. Nevertheless, it is possible a small, 8 bcm/year project for the EEC could start production in the early 1990s.[26]

Qatar's proven gas reserves are massive; estimates vary between 5,660 and 8,500 bcm. The government is particularly interested in developing the reserves in the offshore North field discovered by Shell in 1972. However, cost estimates are astronomical, and the government appears reluctant to commit itself prematurely to building a vast production infrastructure. Development plans submitted by several international oil firms for the North field project include estimates for LNG exports of 8 bcm/year. The IEA believes Qatar's overall export potential is 8 bcm by 1990 and 16 bcm by 2000.[27]

For Iran, under the shah, development of natural gas production was a primary economic objective. Reserves are bountiful—11,000 bcm, 14 percent of the world's total reserves.[28] The EEC was to have benefited from Iranian gas in a triangular deal with the Soviet Union that is discussed later in this chapter. The 1979 revolution, however, not only abruptly ended that project but also put a halt to gas development. Exports ceased for nearly a year and have yet to regain their former volume.

Iran's export potential is phenomenal, but it is impossible to say when, or if, it will come to fruition. In all likelihood, the current dearth of exports to the West will last as long as the Khomeini regime remains in power.

Algeria, already one of western Europe's most important suppliers, has sufficient production and reserve potential to become the EEC's major source of gas. Unfortunately, for political reasons, such a development is dubious at present. The West does not consider Algeria a stable source of supply, a drawback that will limit the size of future contracts.

During the 1970s, Algeria overcame foreign skepticism regarding its ability to finance and oversee large-scale technological projects.[29] But as one of the more vehement supporters of the 1973 oil embargo, it was viewed as an OPEC radical. Despite assuaging statements from government officials to the effect that gas would not be used as a political weapon, the Europeans remained cautious.

In light of the problems that several Western countries have encountered in their gas dealings with Algeria, European reticence seems justified. In 1980, LNG deliveries to the United States and France were halted as part of Algeria's attempt to renegotiate its gas prices to a level of thermal parity with crude oil.[30] The Trans-Mediterranean pipeline, a 1,550-mile engineering marvel that carries gas to Italy via the Sahara Desert and Tunisia, has been a political nightmare for Rome. Gas began to flow in June 1983, but only after the Italian government agreed to subsidize a very high price for the gas. Officials of the Italian national gas company, Ente Nazionale Idrocarburi, claim that the Algerians "blackmailed" Rome by freezing all trade contracts with Italy until the gas deal was consummated. This strong-arm technique reportedly cost Italy $1 billion in trade in 1982. [31]

France signed a contract with Algeria in 1982 to receive 9 bcm/year until 1990, also at a high price, but coupled with Algeria's agreement to purchase French exports. French acquiescence in the hard bargain driven by Algeria was motivated by political ends that took precedence over strict economic prudence—France seeks to maintain and expand cooperation with its former colony. Despite this French deal and the Trans-Mediterranean pipeline, no other EEC country presently wants to be in a position of significant dependence upon Algeria for its gas supply.

The Soviet Union is a candidate to become Europe's largest and most logical future supplier, barring the ideological and political obstacles that stand in the way. In 1950, natural gas made up barely 2.5 percent of the USSR's total energy production. In 1985, with annual production at about 465 bcm, that figure had risen to 27 percent. By the end of this century, natural gas production is likely to surpass that of oil and coal combined.[32] Since Soviet proven reserves are 28 trillion cubic meters (tcm), some 40 percent of world reserves, the USSR is potentially the world's largest exporter of natural gas.[33]

Although gas deposits exist throughout the USSR, most production is east of the Urals, primarily in Siberia. According to the plan for 1980, the European USSR and the Urals were to account for 157 bcm of gas, while regions east of the Urals were to produce 278 bcm.[34] The 1980 plan projected that Siberia alone would supply 155 bcm of gas, equal to almost the entire production for the regions west of the Urals.

Siberia's importance to the development of natural gas production in the Soviet Union extends well into the future. The new regime is calling for total gas production to be increased to 835–850 bcm by 1990. Of that, nearly half will come from Siberia.[35] As one analyst of the Soviet gas industry writes concerning Siberian potential:

> It is not an exaggeration to say that over the next decade, the ability to fulfill plans and produce the desired volumes for domestic and export purposes will stand or fall on the development of this region.[36]

But producing gas in Siberia has proved to be a technological nightmare. This vast region is known for its brutally cold winters and permanently frozen earth, two factors unfavorable to industrial development. Between 1971 and 1975, basic expenditures in Siberia rose 4.8 times, but output increased just 3.6 times. Return on capital fell, and production costs rose.[37] Because of its lack of technological sophistication, the USSR has come to rely on the West for much of the equipment necessary to extract Siberia's riches from the frozen land. The natural gas and oil industries have benefited more than any other Soviet industry from Western technological assistance. In return, the USSR has supplied the West, and especially western Europe, with substantial amounts of its gas needs.

SOVIET GAS TRADE WITH THE WEST

The continuously mediocre performance of the Soviet economy throughout the 1960s and the unwillingness of Soviet leaders to undertake major reforms in the economic system led to a decision to import more Western technology. The Ninth Five-Year Plan (1971–75) called for a 33 percent increase in foreign trade. It emphasized cooperation with Western companies and banks in "working out a number of very important questions associated with the use of the Soviet Union's natural resources, construction of industrial enterprises and the exploration of technical solutions."[38] To finance the additional imports, the Soviets decided to develop the export capacity of their energy sector, notably oil and natural gas. The imports of Western technology that Moscow sought to facilitate would in turn help to increase Soviet energy production.

In 1968, the Soviets concluded an indirect deal with West Germany, using Austria as a go-between to buy German pipe for a domestic pipeline. Under the terms of the contract, two German companies, Thyssen and Mannesmann, supplied the Austrian state-owned firm VOEST with 520,000 tons of 48-inch pipe, which the Russians then purchased from the Austrians.[39]

Two years later, in Essen, the Soviets and the Germans signed the first major "gas for technology" deal of the 1970s. It was to serve as a model for future East-West energy exchanges. The USSR traded its gas for steel pipe, pipe layers, pumps, and compressor stations. Moscow could have purchased the technology directly rather than barter gas for it, but because of the high cost involved and the Soviets' desire to conserve scarce hard currency, it proved more advantageous to couple the import of technology with the export of energy.[40]

Mannesmann, which became the USSR's major supplier of steel pipe, agreed to sell 1.2 million tons of five-foot pipe for $400 million, and to serve as the project's general contractor. Although the terms of the contract, like those of its successors, were not made public, it is known that a consortium of German banks led by Wilhelm Christians' Deutsche Bank provided credits at an interest rate reputed to be 6.23 percent. Mannesmann made up the difference to the consortium between the rate granted to the Soviets and the market rate of 9.42 percent by adding a premium to the price for its equipment. In return, the Soviets were to pump 5.5 bcm/year to West Germany, starting in 1973. A 1,500-mile pipeline stretching from Siberia to Marktredwitz, near the Czech-West German border, was constructed to transport the gas.[41] After Soviet exports had covered equipment and credit costs, the Germans would begin to pay for the gas in cash.

The Germans had strong economic incentives to buy gas from the USSR. The price Moscow offered was attractive to Ruhrgas, the giant German energy company. The steel pipe contracts were a boon to Mannesmann, which had begun to lose a considerable share of the international market upon which its financial well-being depended. German banks, despite the below-market rates of credit, were pleased by the size of the loan, Mannesmann's agreement to make up the difference in interest rates, and the prospect of increased business with the Soviets. Political considerations were apparent as well. The government of Chancellor Willy Brandt, whose goal was to improve and expand relations between East and West, and particularly between the two Germanys, was only too happy to find its aspirations supported by the German business community.[42]

The Germans and the Soviets struck a second deal in 1972. The USSR increased its gas deliveries to 7 bcm/year and purchased an additional 1.2 million tons of steel pipe from Mannesmann, again with the aid of a loan from the German bank consortium at an interest rate of approximately 6 percent, roughly equal to the prevailing market rate.[43] The same parties signed a third contract in 1974 providing for another 2.5 bcm/year of gas. In all, the Soviets were to supply the West Germans with at least 9.5 bcm/year of gas as the contracts came into effect in the late 1970s.

Moscow concluded a number of large-scale gas projects with other EEC countries following the German contracts. By 1977, a total of 11 gas contracts had been signed with western European nations. While the first deals were motivated primarily by market considerations, the OPEC crisis in 1973 prompted the Europeans to look upon Soviet gas as a strategic alternative to Middle East oil.

Despite the generally successful European-Soviet trade relationship, the two deals during this period that received the most attention ultimately failed. The first, Project North Star, called for deliveries of gas from Siberia to the United States. The second involved a trilateral agreement among the Soviet Union, western Europe, and Iran. These ill-fated projects proved to be the inadvertent parents of the Euro-Siberian pipeline.

PROJECT NORTH STAR

During the 1972 Nixon-Brezhnev summit meeting in Moscow, American and Soviet officials discussed the sale of large quantities of Soviet gas to the United States. The communiqué of the 1973 summit meeting in Washington made favorable reference to the prospects of such an arrangement, and the project received a significant boost from the oil crisis that year.[44]

As originally conceived, North Star involved the transfer of gas via pipeline from Urengoy in western Siberia—ultimately the Euro-Siberian's starting point—to Petsamo, on an inlet of the Barents Sea near the Norwegian border. There, the gas was to be liquefied and shipped to the United States at a rate of 20 bcm/year starting in 1980. Most of the gas was to go to the eastern seaboard; it was estimated that the Soviets would satisfy as much as 10 percent of the combined New York, New Jersey, Pennsylvania and New England demand by 1982.[45]

A similar but separate project was initiated at the same time. The Soviets were to pump gas from a field in the Far Eastern republic of Yakutia and sell it in liquefied form to the Japanese and to a California energy consortium for West Coast consumption. Japanese banks granted a $50 million credit to study the plan, but the project was abandoned.

A consortium of American banks was formed to provide below-market-rate financing. In the absence of a nationally controlled gas concern, several private companies, including Tenneco, Tennessee Gas Transmission Co., Texas Eastern Transmission, and Brown and Root, joined together to buy the gas. The cost of the entire project was estimated at $6.7 billion.[46] According to the companies involved, it would generate 250,000 man-years of employment in the United States between 1976 and 2007.[47]

Initially, Washington was enthusiastic about North Star. Secretary of Commerce Peter Peterson stated:

> I believe that these types of joint projects are potentially the single most important product of this new commercial relationship in which the two largest economies of the world each adjust their ways of doing business to the mutual benefit of both. Nor do I expect that these projects will be limited to gas alone, even though the gas projects are likely to be the largest and for that reason deserve high priority.[48]

North Star did not survive the cooling of U.S.-Soviet relations that followed the passage of the Jackson-Vanik and Stevenson amendments in 1974. Without the projected Export-Import Bank loans, government guarantees, and commercial credits that this legislation effectively prohibited, the necessary sources of financing for the deal were unavailable.

During the spring of 1976, an effort was made to revive North Star. Now relying on European financing, the same U.S. gas companies attempted to negotiate an arrangement with the Soviets under which the gas companies would receive 15.5 bcm/year of gas, and France 5 bcm/year.[49] However, the political will to complete the deal had waned in the United States because of the deteriorating political and economic relationship between the United States and the Soviet Union. Congress was skeptical, and certain legislators expressed the fear that the Soviets would cut off the gas supply, or threaten to do so, in order to obtain political concessions. They voiced concern that the project would prove to be a strategic benefit to Moscow because it would expand Soviet energy production. Officials in the Carter administration also felt that pursuing the project might hinder Washington's efforts to improve relations with China.[50] Faced with such opposition, North Star was finally buried and forgotten.

THE TRILATERAL DEAL

As North Star was falling apart, a three-way deal involving the Soviet Union, the EEC, and Iran was concluded in September 1975 at Teheran. The trilateral project called for Iranian gas to be piped into the energy-starved southern regions of the USSR and an equivalent amount to be delivered to western Europe from Siberia via the existing Bratstvof transit system.

The EEC was to receive a total of 10.9 bcm/year of gas, with 5.5 bcm going to West Germany, 3.6 bcm to France, and 1.8 bcm to Austria. In return, the Europeans would cover 80 percent of the gas cost by providing supplies and equipment for the construction of a pipeline between Iran and the Soviet Union. The remaining 20 percent would be paid in hard currency to the USSR. The 20-year contract stipulated that deliveries were to begin in 1981 and reach full capacity by 1983.[51]

The fall of the shah and the rise of Ayatollah Khomeini was the project's death warrant. In addition, an existing contract between Iran and the USSR under which the Iranians supplied the Soviets with gas for their domestic market and for the East bloc satellites was unilaterally suspended by Teheran, causing serious shortages in the southern USSR.[52]

Despite its brief history, the trilateral project did bring Soviet and western European negotiators to the bargaining table again. By 1979, contracts were already in force calling for the Soviet Union to provide West Germany with 15 bcm of gas per year, France with 7.6 bcm, Italy with 7 bcm, and Austria with 4.2 bcm by 1990.[53] Even though the trilateral deal collapsed, the success of these dealings from both the Soviet and the European perspectives made both sides seek further opportunities to expand their gas trade. It was out of this need and spirit of enterprise that the Euro-Siberian was born.

THE SIBERIAN PIPELINE PROJECT

Based on their experience during the early and mid-1970s, the western Europeans had found the Soviets to be reliable suppliers of natural gas.

If the Soviets were willing to sell more gas to the West, they would find the EEC an eager customer. That resource had become an increasingly important part of the EEC's energy diversification program in the aftermath of the first OPEC embargo. Projections showed that indigenous European production could not meet the increasing demand.

In turn, the Soviets had grown to rely upon the West for the technology needed to develop energy production and for the hard currency required to purchase that technology, as well as desperately needed agricultural and industrial products. Consequently, internal Soviet debate on the 1981–85 five-year plan led to a decision to continue heavy investment in the energy sector, with a shift of emphasis from oil to gas.[54] Moscow developed an ambitious plan for the construction of six new pipelines. When the Soviets and Europeans considered Western participation in the project, both sides agreed that in order to increase Soviet export capacity to the West a new line should link the Siberian gas fields directly to the western European grid.[55]

No sooner had the preliminary negotiations begun in 1979, than the second oil shock occurred. By 1980, an OPEC barrel of crude was selling for $31, up from $13 just two years before.[56] Though less substantial in relative terms than the 1973 price hike, this new increase was much more dramatic in absolute terms. Once again, the instigator was OPEC, which had cut production to avert a glut of oil on the world market. Once again, the Europeans felt the blow more heavily than the Americans. Despite successful conservation efforts and increased reliance on other sources of energy, the EEC still depended on OPEC for nearly 50 percent of its energy needs. As the projected purchases of gas from the new pipeline would reduce that dependence to 35 percent by 1990,[57] the Europeans were all the more determined to see the project through to the end.

Upon his return from a state visit to Moscow in July 1980, West German Chancellor Helmut Schmidt officially announced the project. Originally, the Soviets planned the Euro-Siberian pipeline to run just over 3,000 miles from Yamburg, which is north of the Arctic Circle, down to Urengoy, and then across much of the Soviet Union to Uzhgorod near the Soviet-Czechoslovak border. From there, it would be patched into transfer stations at the West German-Czech and Austrian-Czech borders, and gas would be distributed to the contracting EEC nations. At full capacity, a total of 40 bcm per year was to flow to the West.[58]

By 1980, however, the focus of the Euro-Siberian had shifted to Urengoy. Yamburg did not possess a sufficient infrastructure to allow the rapid development of its gas deposits; constructing that infrastructure would be an additional cost and take too much time. Besides, Yamburg's climate was especially inhospitable. The starting point of the pipeline was therefore moved to Urengoy, the largest single natural gas complex in the world with about 7 bcm of proven reserves—20 percent of Soviet and 7 percent of the world total.[59] From there, the Soviets would build, with Western assistance, a 4,465-kilometer pipeline across vast stretches of permafrost, the Ural mountains, seemingly endless swamps, and numerous rivers and streams. As an engineering and construction task alone, the project seemed herculean, but conflicting American and European policies proved far more complex than the necessary miracles of technology.

In addition to decreasing European dependence on OPEC, the Euro-Siberian pipeline promised to be a much-needed boon to Western industry stricken by a deepening worldwide recession. Even healthy economies like that of West Germany were beginning to suffer from high unemployment. By 1982, nearly 2 million people—7.5 percent of the work force—were without jobs in the Federal Republic. In France, 8 percent were unemployed, in Italy 9 percent, and in Great Britain, the rate reached 12 percent (nearly 3 million people). Newspapers, commentators, and political figures noted that the last economic crisis of such magnitude had preceded and, some might argue, had led to World War II. European leaders feared that a few more percentage points on the unemployment scale would push domestic instability beyond their control.

Within the individual EEC countries, large-scale industry was especially hard-hit. For many of these companies, the equipment orders for the proposed pipeline were lifesavers. Mannesmann, the giant pipe manufacturer that had played such an important role in the earlier German-Soviet gas projects, had $52 million in projected losses for 1980. AEG-Kanis, a subsidiary of AEG-Telefunken that manufactured compressor stations, had not made a profit since 1976, and the parent company itself was on the verge of collapse. Both Mannesmann and AEG-Kanis had had to absorb the cancellation of the trilateral deal and some smaller projects with Algeria, the general economic recession, and increasingly effective competition from the Japanese. For Mannesmann, whose large-diameter steel pipe exports to the USSR constituted 60 percent of its production and guaranteed 2,500 jobs, the new pipeline would provide at least 1,000 additional jobs. According to AEG-Kanis chairman Heinz Durr, the Euro-Siberian would assure his company "20,000 to 25,000 jobs over a three year period" and perhaps prevent the failure of AEG-Telefunken.[60]

Elsewhere in the EEC, the suppliers of steel pipe, compressor stations, pumps, and other energy-related equipment were looking to Siberia for help in weathering the recession. In 1980, Creusot-Loire, a major French pipe producer, appeared to be doomed. Between 1977 and 1980, it lost 16 billion French francs, roughly $2.5 billion in 1980 dollars.[61] Creusot ultimately collapsed in 1985. The Italian pipe maker Finsider was counting heavily on Soviet orders for its product; otherwise, it would have to begin laying off personnel. Similarly, Nuovo Pignone in Italy and John Brown and Co. in Scotland, both manufacturers of compressor stations, desperately needed the Soviets' business.[62] For all of these companies, there were no other markets comparable in size with that of the Soviet Union.

The French and German governments were unwavering in their support for the pipeline. In part, this reflected a heightened sense of accountability on the part of each government for the ongoing recession. The unemployed in Europe typically hold the state responsible for their predicament, more so, certainly, than the jobless in the United States. They are quick to blame the government for hardships resulting from economic dislocation, even when the problems at hand are beyond a quick fix. Even marginal intervention to curtail exports requires extremely persuasive justification. Cancelling participation in the pipeline could have entailed serious political consequences, the governing coalitions in France and Germany believed, all the more so because elections were approaching in both countries. In short, the jobs the pipeline promised had become almost a political necessity. By guaranteeing the financial aid packages negotiated by their banks with Moscow, the French and West German governments could show their constituents that they were taking an active role in fighting the recession and creating employment.[63]

Thus, business leaders and government officials in western Europe began to see in the proposed pipeline a timely, and even crucial, injection of life into their stagnant industries. This served to secure hearty endorsements for a project that would also provide the EEC with a badly needed and inexpensive resource free of the whims of the Middle East. Nor could the geopolitical realities be ignored, for while détente was not dead in the West, it was clearly ailing. The

Euro-Siberian would be a physical, living link between East and West at a time when relations between the two blocs were fast reverting to Cold War status in the face of intensified hostility in Washington from the Reagan administration and international disapproval of the Soviet invasion of Afghanistan.

3

The Contracts

Since the Euro-Siberian pipeline would be the largest East-West commercial venture ever, it quickly became clear that the project was more than just another gas deal. The scale and difficulty of the undertaking were immense. A 3,000-mile pipeline spanning some of the most inhospitable terrain in the world would require billions of dollars in sophisticated technology and steel pipe.

A complex set of negotiations in three distinct areas began. European bankers, leaders of industry, and government officials attempted simultaneously to arrange the terms of financing, equipment sales, and gas prices with the Soviets. The provisions of the Euro-Siberian contracts are secret. Some of the data and figures presented here have been obtained from clauses and excerpts examined through the kindness of certain persons with access to the documents. Much, however, is based on press reports, academic studies, and interviews with businessmen, bankers, and government officials directly involved in the project.

Some critics of the pipeline believe that the project was a poor business undertaking for the West. They claim that the credit arrangements were too generous and the price of Soviet gas too high, and they argue that Western businessmen were falling over themselves to accept contractual clauses favorable to Moscow in their eagerness to participate in the deal. Furthermore, these critics warn that long-term shifts in the world energy market will make the Soviet gas unnecessary. Were western Europe's proverbially shrewd businessmen and bankers outnegotiated by the Kremlin? Was the Euro-Siberian pipeline the deal of the century, or a monumental economic blunder?

CREDIT ARRANGEMENTS

In early 1980, as negotiations for credit arrangements began, the project appeared destined to follow the model established in the 1970s by the first East-West energy deals: European banks would form a single consortium to provide loans to the Soviets. Six months later, however, Moscow abruptly changed its mind and insisted on separate negotiations with individual European banks or single-nation consortia.

The Soviets asked for, and initially received, three times the amount of funding they needed to purchase Western equipment. The Germans, headed by the Deutsche Bank, offered $5.2 billion. The French, led by the Credit Lyonnais, granted $3.3 billion. A Belgian group directed by the Société Générale de Banque contributed $1 billion, and Holland's Algemene Bank Nederland, $1.25 billion. At the same time, the Soviets negotiated a $3 billion credit with the Japanese Export-Import Bank and discussed loans with the British, Italians, and Austrians.[1]

There has been much speculation as to why the Soviets sought an overabundance of credit. In an extremely thorough analysis of the negotiation process, Thane Gustafson of the Rand Corporation suggests two possibilities. First, the Soviets may have anticipated difficulties in obtaining large volumes of credit from individual countries and were padding their requests accordingly. Second, Moscow may have thought about using the excess credit to purchase Western equipment for the expansion of its domestic supply lines.[2] Both explanations are certainly plausible.

Discussions centering on the applicable interest rates were also proceeding in a highly competitive environment. The French agreed to a rate of 7.8 percent, consistent with a Franco-Soviet commercial protocol signed in January 1980 and due to expire in the summer of 1981. The OECD had established a similar lending rate for certain nonmember countries. However, interest rates were increasing dramatically at that time. In 1979, they had averaged 9.6 percent in France, reached 12.96 percent by the first quarter of 1980, and were up to 17.46 percent by the third quarter of 1981.[3]

Although the French government would grant export subsidies in order to make up the difference between the low interest rate charged the Soviets and the market rate, German bankers had traditionally refused such support. Deutsche Bank officials were unhappy about the favorable terms the Soviets had obtained from the French and then used as a bargaining chip in their negotiations with other banks.[4] Nevertheless, the German interest rate remained consistently lower than that of its allies: 6.65 percent in 1979, 9.08 percent in the first quarter of 1980, and 12.76 percent in the third quarter of 1981.[5] The difference between the rates granted to the Soviets and the prevailing market rate was, at least initially, negligible. And, as we have seen, the German banks usually recouped the difference from equipment suppliers who charged the Soviets a premium on their products of up to 20 percent.

But then the negotiations broke down. The Europeans refused to grant more than 85 percent of the loans at below-market rates. That was the maximum that their government export insurance agencies, such as Hermes in Germany and COFACE in France, would cover. The Soviets wanted 100 percent. The impasse lasted through the first months of 1981, at which time the Soviets finally realized that skyrocketing interest rates in western Europe were eroding their bargaining position. In addition, the parallel equipment negotiations were nearing an end, and could not be closed until credit was available. By this time, however, the French and the Germans had withdrawn their original financing offers, and the Japanese had delayed approval of their Export-Import Bank loan.

A new set of financial arrangements was finally concluded in July 1982, under which the Soviets agreed to 85 percent coverage and consented to repay the loans over seven to eight years instead of ten. The Germans ultimately provided about $2.5 billion in credit, and the French $2 billion. Moscow signed smaller standby commitments with the Japanese, the Dutch, the British, and the Belgians that would take effect if costs proved greater than anticipated, but most of this credit went unused.[6] In all, the financing package came to between $5 and $6 billion.[7]

In retrospect, it is clear that the Soviet Union had a strong bargaining position for obtaining the credit terms accorded for the purchase of Western equipment and technology. Moscow was in effect the sole buyer of a product being offered by a multitude of competing suppliers. Gordon Crovitz has written that "Soviet negotiators were careful to take advantage of the Soviet Union's position as a single negotiator faced with several competing traders. In the case of the pipeline, the Soviet Union was a monopsonist . . . [and thus able to] divide and conquer."[8]

Axel Lebahn, a Deutsche Bank official involved in putting together the largest financial package for the pipeline, describes the negotiations in this manner:

> There was a kind of negotiation delegation, as a rule, first visiting the German partner, then going the rounds of interested parties in Italy, France, etc. . . . In this way, the negotiations were conducted virtually in series, and step by step, with the competing countries and firms. This enabled the Soviet delegation to present the lowest Western bid for any part of the contract almost simultaneously to all Western competitors as the most the Soviets could possibly consider. As talks progressed, the Soviet Union skillfully played the various countries off against each other; at a later stage of negotiations it succeeded in negotiating the best possible terms with a growing number of competitors within Western countries.[9]

Deutsche Bank president Wilhelm Christians echoed Lebahn's description: "It is true we would have preferred tougher terms. The Soviets were negotiating from strength."[10] This aspect of the pipeline project was particularly irksome to critics of East-West trade, who believe that, even if they cannot put a stop to commercial relations between the two blocs, they can at least impose stricter and more prudent economic terms. "Our aim should be to bring our dealings with the Soviet empire back within the jurisdiction of the laws of the marketplace," wrote French philosopher and author Jean-Francois Revel in a critique of the pipeline. "We must force the Soviet Union to face economic facts by offering it trade not aid. . . ."[11]

The Soviets clearly did well in negotiating the terms of credit. They were granted interest rates of 7.8 to 8.76 percent at a time when market rates stood at around 12 percent in West Germany and 15 percent in France. However, these terms were consistent with the Franco-Soviet commercial protocol then in effect, and also with the OECD Consensus governing lending rates applicable to nonmember states, which had not yet been revised. Moreover, countries give generous and often subsidized financing terms to promote their exports, regardless of destination. The terms should not be compared with those prevalent in a normal economic environment, but rather with those found in the export trade, especially for countries such as EEC member states that follow an "export or perish" philosophy. These nations are not geared to strict economic prudence in their dealings with other countries. France, to cite one example, finances a large volume of exports to Third World countries where the chance that the principal will be repaid is small.

The fact that the Soviets won very good terms even by this standard is not so unusual. As the first energy agreements between western Europe and the USSR show, such favorable terms of credit are common. The West typically provides large sums of credit—often several billion dollars—at below-market interest rates with lengthy repayment schedules. There is a sound economic justification for these arrangements. Fiat chairman Giovanni Agnelli believes that "Even if intense inter-Western competition, against a monopolistic buyer, may sometimes reduce profit margins to narrow limits, the very size of the 'buyer' as well as the size of the deal this gigantic buyer may offer, make these relations rather unusual."[12] Following the 1973 oil crisis, western European governments adopted aggressive export policies in order to support their industries and balance their growing trade deficits.

In the past, Western businessmen have also been able to add a premium to the price of the products sold to the Soviets as a quasi setoff for low interest rates. According to Thane Gustafson, the Soviets go along with this because

> The interest rate on export credits is a highly visible indicator, and the rate struck in one deal immediately becomes the standard for all subsequent negotiations. The performance of the Soviet Bank for Foreign Trade [Vneshtorgbank] is presumably measured in part by its success in controlling this key number. In contrast, the purchase price of industrial equipment is a more elusive quantity, since broadly similar items may differ in their specifications and hence in their prices. Mashinoimport and the other Soviet purchasing organizations are presumably judged by a host of indicators, of which price is only one.[13]

In this regard, it is not clear how closely the Siberian pipeline contracts followed the model. As the Deutsche Bank's Christians remarked, the pipeline project was more complex than its predecessors because of its multilateral nature, which was the key to the Soviets' negotiating advantage. Some of the prices charged for pipeline equipment were considerably lower than usual. Western companies, desperate for business during the recession, were more willing to cut profit margins. Christians hints that the German pipeline contracts, unlike those signed by the French and the Italians, may have been more favorable to German business interests. Apparently, some sort of "triangle deal" involving the terms of credit, price of equipment, and gas price was worked out.[14] There is no hard evidence to support this claim, but such an arrangement certainly could have been facilitated by the fact that Christians, whose bank organized entirely private credit for the pipeline, sits on the board of directors of Mannesmann, the steel pipe manufacturer.

EQUIPMENT SALES

The Soviets employed the same strategy in negotiating the equipment contracts that they had used in the financing arrangements. In the past, they had hired a single contractor—usually Mannesmann—that then found the necessary subcontractors. For the Euro-Siberian pipeline, however, Moscow decided to bargain directly with the subcontractors in order to obtain more favorable prices. The Soviets set up a negotiating team in Cologne and let the European manufacturers come to them. This tactic, designed to play one party against another in order to win the best terms, was well established in other areas of East-West trade. Often, any advantages Moscow had seemingly achieved were in reality offset when Western companies padded their bids and sold the Soviets outdated technology. By most accounts, though, the tactic worked well in the pipeline bargaining: Moscow reportedly extracted a discount of almost 20 percent on the most costly items—the 125 turbines needed for the pipeline's 41 compressor stations.[15]

The first contracts were signed in September 1981. Germany's Mannesmann Anlagenbau, a subsidiary of Mannesmann, and France's Creusot-Loire would deliver and install 21 compressor stations for just under $1 billion. They also agreed to oversee the work of two subcontractors, AEG-Kanis and John Brown and Co., that were to supply 68 compressor turbines for approximately $500 million. As compensation, Mannesmann and Creusot-Loire would receive a fee of 6 percent of the total equipment cost. Mannesmann alone would provide $2 billion in steel pipe and about $60 million worth of compressor equipment such as piping systems, skids, and pumps.[16]

By December, the other major equipment contracts had been concluded. Moscow awarded Nuovo Pignone the contract for the remaining 57 compressor turbines, valued at roughly $360 million. In addition, for a 6 percent fee, the Italian company agreed to oversee all work on the southern portion of the pipeline. Thomson-CSF, a leading French electronic and computer firm, won a bid to furnish more than

$300 million worth of sophisticated monitoring equipment. Another French company, Alsthom-Atlantique, undertook to supply 40 sets of spare rotors for the pipeline's principal turbines.[17]

Several American companies, notably General Electric, Caterpillar Tractors, Dresser Industries, and Cooper Industries, also took part in the transaction. GE, the world's leading manufacturer of turbines and related equipment, was to sell the Soviets $175 million in rotor kits through its European licensees. Caterpillar would provide 200 pipe layers worth $90 million, most of which had already been contracted for by the Soviets for other projects. All in all, the value of the American participation in the pipeline was estimated at $300 to $600 million.[18] However, as a result of President Reagan's initial ban on the sale of pipeline equipment in December 1981, many of these contracts never took effect.

Criticism of the equipment sales centers on a number of contractual clauses. For example, some have pointed to a damages provision that allows the Soviets to collect up to 5 percent of the contracted equipment's value for nonfulfillment of supply obligations. In fact, such clauses are a standard feature of Soviet contract practice because of the USSR's centrally planned economy. Every requirement must be fulfilled on time, or the entire plan will be disrupted. In the Soviet system—in contrast to the West—nonfulfillment of a contractual delivery obligation cannot be "made whole" by money damages. The Soviets insist upon specific performance. Hence the severe contractual penalty provisions, which are designed to deter suppliers from reneging on an agreement.[19]

Another criticism concerns a clause that was removed from the equipment contracts. Most international trade agreements include a provision absolving the supplier of responsibility for delivery in the event of unavoidable and unforeseeable circumstances. This "force majeure" clause usually covers such natural disasters as floods, typhoons, earthquakes, and epidemics. It also typically takes into account government intervention such as export prohibitions. This last contingency was edited out of the Euro-Siberian equipment contracts at the behest of the Soviets. As a result, the Europeans would have been obliged to pay the penalty price had they failed to deliver the goods when Reagan declared his embargo. As one veteran of East-West trade negotiations notes, the Soviets always try to "white out" this part of the force majeure clause, while Western lawyers try to keep it in. Usually, the Soviets prevail.[20] This is certainly evidence of Moscow's astute understanding of international trade subtleties and political realities, but it hardly seems grounds for condemning the pipeline.

GAS PRICING

When negotiations on the terms under which the Soviets would supply their gas began in 1980, it was thought the total volume would be between 38.5 and 40.5 bcm/year of gas to all the participating EEC countries. West Germany would take 10.5 bcm/year, France and Italy 8 bcm/year each, Belgium 5 bcm/year, Austria 3 to 5 bcm/year, the Netherlands 3 to 4 bcm/year, and Switzerland 1 bcm/year.

By 1982, though, several European countries had decided to reduce or cancel their participation in the project. These reversals have no uniform explanation. The Dutch were not satisfied with Soviet pricing demands and, buoyed by their own substantial domestic gas production, withdrew.[21] The Belgians, displeased because none of their companies had been offered equipment contracts by the Soviets, also backed out. The Austrians purchased less gas than expected because the recession had led to a projected decrease in natural gas consumption of 5 to 6 percent for 1982. Without equipment sales for the pipeline, they also lacked the income necessary to offset the cost of the Soviet gas.[22]

Five governments finally ratified the gas contracts negotiated by their national gas companies, with the Italians holding out until June 1984. Under the contracts as signed, the Soviets will supply West Germany with 10.5 bcm/year, France with 6.4 to 8 bcm/year, Austria 1.5 bcm/year with an option for an additional bcm/year, Switzerland 0.36 bcm/year, and Italy 6 bcm/year. The Euro-Siberian will thus carry 24.76 to 27.36 bcm/year of gas to the EEC by the time supply flow reaches full capacity in the 1990s. Coupled with all previous contracts, the Soviets will sell the western Europeans 57.16 to 59.76 bcm/year of natural gas.

Establishing a price for the Soviet gas proved to be one of the most complicated aspects of the pipeline deal. Unlike oil, whose price is generally set by its exporters or by the Rotterdam spot market, gas has no unified market price. Trade terms are negotiated whenever a new source becomes available, and usually are renegotiated at regular intervals to take into account changing energy market conditions. Moreover, pricing agreements often include some kind of index of competing energy sources to which the value of gas is keyed. For these reasons, there are considerable discrepancies in the price of gas throughout the world.

The European and Soviet negotiators agreed that the price of Siberian gas would equal the sum of a base price plus a weighted index of reference fuels. The Soviets wanted a base of $5.50 per million British thermal units (BTU), and an index of 100 percent parity with crude oil. The Europeans considered both demands unreasonable. At the time, Dutch gas sold for about $4.20 per million BTU, and no exporter had ever been granted 100 percent parity with crude as a pricing index. Consequently, the Soviets abandoned their 100 percent parity demand by the fall of 1981, dropping it to 50 percent, and finally to 20 percent. This yielded a definitive weighted reference fuels index of 20 percent crude, 40 percent heavy, and 40 percent light fuel oil.[23]

The base price was also reduced significantly. In late November, the Soviets and a German consortium composed of Ruhrgas-AG, Thyssengas, Gelsenberg (BP), Salzgetter-Ferngas, and Gewerkschaft Brigitta-Elwerath settled on a base of $4.70 to $4.80 per million BTU.[24] Shortly thereafter, the French achieved a base of $4.65 and the Italians one of $4.70. Because the Italian government delayed ratification of the contract, that price was renegotiated in 1984 down to $3.60 per million BTU. The decrease reflected a glut on the world energy market.[25] In order to protect the Europeans from an overvalued dollar, the prices were quoted in the currency of each country, and payment was allowed in deutsche marks, French francs, and Italian lire.[26]

The contracts called for a floor price, or guaranteed minimum, to ensure Moscow sufficient gas revenues in the event of a sharp decrease in the price of oil. The Soviets initially asked for a floor price of $6.05 per million BTU. Ultimately, the Germans agreed to a minimum price of $5.70 as of 1989, and the French a minimum of $5.50 commencing in 1984.[27] If the market-related price dropped below the floor price, the latter would prevail until the purchasers had "overpaid" a sum equal to about 10 percent over the value of a year's deliveries at the market-related price. At that point, the market-related price would take effect.[28]

The floor price would currently make Siberian gas overpriced in comparison with oil. If the original contracts were now in effect, the French would be paying $5.70 per million BTU and the Germans $5.90 per million BTU, the equivalent of $34 to $35 for a barrel of oil. The price of an OPEC barrel ranged from about $26 to $28 in 1985, and dropped as low as $10 in 1986.

However, in December 1984, the Soviets agreed to suspend the floor price.[29] They then entered into negotiations to revise the contracts, under which the Europeans can renegotiate the price and volume of gas every three years. By the summer of 1985, both the French and the Germans had succeeded in substantially lowering the price of Soviet gas to between $3.50 and $3.80 per million BTU. Italy had already won a similar price by delaying the ratification of its gas contracts until 1984. The Soviets also were willing to decrease the volume of their sales to the EEC. The French had contracted for a flow of 8 bcm by 1986; now, that flow will be achieved only in 1990. This means a savings of over 1 billion francs. Finally, the indexing formula used to calculate the price of gas was also modified. The factor of gas oil was increased, and a factor for coal was added. In the long run, this will help ensure that the price of gas remains in line with that of competing fuels.[30]

Such flexibility on the part of the Soviets may seem surprising. In fact, it made good economic sense from Moscow's vantage point. Soviet strategy was to cut prices, and even volumes in the short term, in order to capture as much of the long-term European gas market as possible. Taking into account all the contracts with various suppliers in effect prior to the North Sea Troll project, the IEA projected that 10 percent of the EEC market would still be available in 1995.[31] The Soviets wanted to be in good standing with the Europeans to win as much of that remaining share as possible. They would happily undercut other potential suppliers on price to do so. However, the Troll development is now expected to meet a large part of the outstanding EEC demand. The Europeans are thus in the enviable position of having won substantial price reductions from Moscow while still holding the Norwegian gas bargaining chip with which they might negotiate additional reductions in the future.

Whatever the long-term results, the Europeans are pleased with the gas supply contracts because the renegotiations were so successful. Of course, not everyone is satisfied. Some fear that the burden of a lower price would be borne not by the Soviets, but by European taxpayers. Assuming that Moscow budgeted its loans upon the basis of the original contracts, the change in price could derail the Soviet service and repayment schedule. Should the Soviets be obliged to default, the EEC governments would in all likelihood be obligated to honor their guarantees to the relevant financial institutions.[32] This scenario seems farfetched, however, since the Soviet balance of trade with the West is sound and debt has been reduced.

The European bankers do not seem worried about Moscow rescheduling or defaulting on its payments. "In my experience of arranging deals for 16 years, I wouldn't hesitate to say the Soviets are a good credit risk," claims Christians. "They are very concerned about their image as a reliable business partner."[33] This is an imperative dictated by the nature of the Soviet economic system. In the West, if one company defaults or otherwise angers a trading partner, its conduct does not rub off on other companies, which are independent of the wrongdoer. In fact, it probably helps them by decreasing competition. If the Soviets default on any given deal, however, the entire centrally controlled system looks bad. Moscow is thus conservative when it comes to managing its debts.

Another criticism of the pipeline contracts concerns the minimum amount of gas the Europeans must purchase, regardless of world energy market conditions. A clause obligates them to "take or pay" for 80 percent of the contracted volume of gas. This provision would seem especially troublesome in the event of a glut on the market, should the Europeans discover that they have contracted for too much gas. In fact, the gas market is now experiencing such a glut.

However, there is another clause in the contracts that allows the Europeans to renegotiate annual liftings upward or downward by 20 percent at the start of each year. In glut conditions, the EEC buyers could reduce deliveries by 20 percent, and then take or pay for 80 percent of the remainder. For example, the French originally contracted to take 8 bcm/year. At the start of a given year, they could lower that to 6.4 bcm, and then be required to purchase only 80 percent of that, or 5.1 bcm/year.[34] And, of course, the Europeans can renegotiate the base volume of gas they must take every three years. Despite "take or pay," the gas contracts are quite flexible.

"Take or pay" is a common practice in the international gas trade, not some clever Soviet invention that the Europeans too quickly accepted. In fact, the "take or pay" clause in other EEC gas contracts is much less favorable to the Europeans, reaching 90 to 95 percent for

Algerian gas.[35] In addition, the "pay" part of the clause apparently allows an EEC country to purchase the gas, but requires the Soviets to defer delivery until a later date determined by the customer, thus averting storage costs.[36] While this report is unconfirmed, the Soviets would probably welcome such an arrangement, since they would receive payment in advance of delivery.

THE CHANGING WORLD

Since the pipeline contracts were signed, European energy consumption has decreased substantially. West Germany now consumes about 8 percent less energy than it did in 1979, and France nearly 10 percent less.[37] In addition, Norway's decision to vastly increase exports to the EEC has created, at least potentially, an overabundance. This has helped produce dramatic changes in the world oil and gas markets. Critics of the Siberian pipeline argue that the prevailing glut conditions are proof the EEC will not need much of the Soviet gas it must purchase. Some even claim the pipeline may have been unnecessary.

When the pipeline contracts were first negotiated, the 1979 oil crisis had pushed the price of a barrel of oil to $34. The subsequent glut, combined with division within OPEC, caused the price to fall to about $26/barrel. By the beginning of 1986, spot market prices had tumbled as low as $10/barrel, and the Western press printed OPEC obituaries. Such optimism may be ill-founded. While some experts see a continued, irreversible erosion of oil prices, others believe that the price will continue to fluctuate, and ultimately will rise well above its $34 high. The International Energy Workshop, in a poll of forecasters from around the world, now projects $40/barrel in 2000 and $60/barrel in 2010, using 1984 dollars as a base.[38]

Numerous factors contributed to a drop in oil consumption over the past few years that pushed the price down. Just how many of these will induce permanent savings—such as home insulation and smaller cars—and how many represent temporary economies—such as driving less and setting thermostats low—is difficult to know. Moreover, the greater the economic recovery in the West from the recession of the late 1970s and early 1980s, the more energy will be consumed. This could help push the price of oil upward again.

The great irony is that the oil price collapse was also provoked by diversification efforts in the West that included the Siberian pipeline.

By successfully replacing oil with natural gas, coal, and nuclear power, the Western nations redefined the supply-demand equation that had favored OPEC throughout the 1970s. Since the pipeline helped produce lower oil prices, however temporary they may be, it is somewhat disingenuous to argue that Soviet gas is useless to the Europeans because of the oil glut.

Gas consumption has also decreased since the pipeline contracts were drawn. The IEA's 1982 projection that its member states would use 316 bcm of gas/year by 1990 has been revised to between 250 and 270 bcm. Even short-term estimates have proven faulty: the IEA nations consumed 9.5 percent less gas in 1983 than had been predicted the year before.[39] Here, too, the causes for the gas glut were the success of energy conservation measures and the lingering economic recession.

If the Europeans did overestimate their need for Soviet gas, the energy crisis of that period at least makes their miscalculation understandable. The flexibility of the contracts has already allowed the EEC to reduce purchase volumes and prices. Moreover, the long-term outlook for the gas market in particular, and the energy market in general, is highly uncertain. Bitter past experience dictates that it is better to have too much rather than too little energy on hand.

The most recent IEA report finds that "Even concerted policy actions by the OECD governments to accelerate their own gas development will not eliminate the need for growing quantities of imported gas." For the year 2000, the IEA now projects that the OECD's demand will fall between 310 and 384 bcm/year, while indigenous production will be between 125 and 181 bcm/year. That leaves room for between 185 and 203 bcm of imported gas.[40] Even with the Euro-Siberian, the USSR will supply only 60 bcm of European imports.

If the Europeans discovered anything during the oil crises of the 1970s, it was that energy supply diversification had to become a policy priority. That imperative remains. At present consumption rates, the world has only another 60 years of oil left.[41] Short-term gluts can be an inconvenient side effect of diversification. But that does not suffice to remove a central justification for the Siberian pipeline. As sovietologist Marshall Goldman concludes, "You find energy wherever you can and build redundancy wherever you can."[42]

The economic critique of the Siberian pipeline proved to be a sideshow in comparison with the debate that arose over the project's strategic implications. Washington's anti-pipeline case centered on the alleged threat it posed to Western security. Had western Europe made itself too dependent on the Soviets for energy? How significant a hard currency windfall would the Soviets achieve from the additional gas sales? Would the technology sold to the USSR for the pipeline prove to be a boon to the Soviet economy? These were the questions at the core of the alliance crisis.

4

Strategic Issues

"If I were a Soviet leader, I would have rubbed my hands with delight when the Europeans signed the pipeline contracts," says a senior official in the U.S. Department of Defense. "Can you imagine getting a primary adversary in the position of depending on you for energy, with both of you knowing that the tap can be turned off at any time? That's the kind of leverage strategists usually only dream about."[1] From Washington's perspective, the Europeans had put themselves in an uncomfortable position by contracting to purchase so much Soviet gas. The administration was also concerned about Moscow's potential hard currency earnings from the pipeline and the impact of Western technology on the Soviet energy sector. These strategic critiques were the core of the United States' case against the Euro-Siberian.

It is a natural political reflex in the West to scrutinize any project eagerly pursued by the Soviet Union. Prudence renders such an investigation imperative. Ultimately, fears concerning the pipeline's strategic drawbacks proved to have been greatly overstated. Nonetheless, the project's critics raised important and legitimate points that deserve careful analysis.

There is an ethical question as well. Although little concrete evidence exists, it is likely that the Soviets employed forced labor to build the pipeline. If that is true, should this abuse of human rights have prompted the West to abandon the project?

SOVIET ENERGY LEVERAGE

The main concern about the pipeline was that it would make western

Europe too dependent on the Soviet Union for gas. This raises two separate questions. First, what would be the consequences if the Soviets abruptly cut off the supply of gas to the EEC? Second, what is the likelihood that Moscow would do so?

In a quantitative sense, the degree of western Europe's dependence upon Soviet gas is clear enough. By the time the pipeline reaches full capacity in the early 1990s, the Soviets will supply up to 30 percent of the gas needs of France and West Germany, or some 6 percent of their total energy consumption. Soviet gas will constitute up to 11 percent of Italy's overall energy consumption, and as much as 40 percent of Austria's.

However, efforts to assess the significance of these figures have caused considerable debate in the West. At first glance, it would appear that depending on one supplier for no more than 6 percent of one's total energy needs does not represent a grave danger. Certainly, few people objected to the fact that a schizophrenic Libya provided West Germany with precisely that percentage of its primary energy supply in 1981. Averages, though, can be misleading. Richard Pipes, the former national security adviser for Soviet affairs, proposed this analogy:

> Say you have flooding in one half of a country, and below normal precipitation in the other. The average would make conditions for the entire country seem normal. But one half is suffering very much. With Soviet gas, we have a similar phenomenon. You must look at regions and cities, not all of Europe or even an individual country.[2]

In short, the "end use" of gas must be analyzed. This poses some difficulties. While it is possible to ascertain the gas consumption of a specific economic sector or individual region, it is often impossible to learn where that gas originated. In West Germany, for example, a sophisticated gas grid mixes the Soviet supply with other imported as well as domestic gas, so that by the time the Soviet gas reaches its ultimate destination, it cannot be distinguished from other sources.[3]

We do know that certain regions will receive an inordinately high percentage of their gas from the Soviet Union. Bavaria, home base for the West German automobile and chemical industries, will rely on Soviet sources for as much as 95 percent of its gas, as will West Berlin. These, and a small number of similar cases, could potentially be a cause for concern.

The end use of gas by sector tells us more. Gas is consumed primarily in the residential and commercial sectors; industrial use is less important, and gas is insignificant as a source for electricity. The IEA projects that by the year 2000, gas will account for 39.8 percent of the energy needed by the EEC residential and commercial sectors, 25.2 percent of the energy required by the industrial sector, and only 1.5 percent of the energy used for the generation of electricity.[4] Soviet gas will represent no more than a third of the percentage of gas for each sector. In a sector-by-sector analysis, the importance of Soviet gas is brought into sharp focus. Strategically, an energy loss in the commercial and residential sectors is far less serious than in the industrial and electricity sectors, which are vital to a country's defense needs. Generally, a cut in, or even an end to, the flow of Soviet gas to the EEC would have limited adverse strategic effects.

The Europeans possess certain safeguards that would diminish the impact of a gas crisis. Some sectors have dual-firing capability, which means they can switch from one energy source, such as gas, to another, such as oil, in a short period of time. Some 75 percent of West German gas-using utilities can convert rapidly to oil.[5] In the event of a "double crisis" involving both oil and gas shortages, dual-firing capability would be a less effective safeguard. While it is true that the commercial and residential sectors in the EEC have not developed a dual-firing capability, this does not pose an important strategic problem.

Gas grid flexibility is another means to offset a sudden loss of supply. The German grid includes an emergency rationing system specifically geared to deal with a supply stoppage by any of the Federal Republic's foreign suppliers. In Austria, the Soviet pipeline enters the country at Baumgarten, the site of a major gas field. This would allow the Austrians to make up for a cut in the Soviet supply with their own gas.[6] Production at the Groningen field in the Netherlands also could be raised on short notice in a crisis, thereby partially alleviating gas shortages within the community.[7] The EEC grid system, connecting most of the member states except Great Britain, Sweden, and Spain, permits gas to be moved easily from one country to another at short notice. A supply cut directed at a particular nation could thus be alleviated. The grid also makes it difficult for the Soviets to deny gas to one country without implicitly threatening all the others linked to the grid. Gas leverage would be unwieldy.

Additional steps would further secure western Europe's energy supply. A switch from gas/oil to gas/coal dual firing would remove the threat of simultaneous gas and oil shortages. Storage capacity should be increased. At present, the French, Germans, and Italians store enough gas for 35 days of use at a constant consumption level.[8] The greater this storage capacity, the larger the supply of gas the Europeans would have at their disposal in a crisis.

In sum, a disruption in the supply of Soviet gas could have significant negative effects upon a small number of individual regions within the EEC, and also upon the commercial and residential sectors. However, this assumes a nearly total stoppage in supply. On the whole, a cutoff would be far from catastrophic and, according to most studies of the question, poses no long-term threat either to EEC economies or to the Community's defense capacity. As one expert concluded, "A cut-off in Soviet gas, while clearly troublesome, would be far from a disaster for Western Europe."[9]

In spite of these findings, some strategic planners remain concerned that even a short-term disruption poses a real threat to Western security. They argue that the inconvenience to the EEC of a threatened cutoff could suffice to prompt government officials to accede to Soviet demands on a given political issue, especially if it were accompanied by the simultaneous threat of an oil embargo from the Middle East. They also fear that in a military confrontation between East and West, a disruption in Soviet supply could give Moscow a decisive short-term tactical advantage. Hence the continued relevance of the question of the probability of the USSR's ever using such gas leverage as it possesses over the EEC for political benefit, or in the event of a military clash.

Opponents of the pipeline point to the Soviets' use of the oil weapon in the past and suggest that Moscow would not hesitate to utilize its gas exports in the same way. In 1948, the USSR denied oil to Yugoslavia when that country sought to break away from the Soviet bloc; Albania lost its Soviet oil in 1961 for aligning itself with China; the Chinese were forced to do without Soviet oil when the two countries split over ideology and economic strategy in 1962; the Soviets canceled an agreement to supply oil to Israel's Haifa refineries following the 1956 Suez crisis; and they reduced or cut off their supply to several Third World countries that shifted allegiance to the West or otherwise deviated from the Soviet line.[10] The interesting feature of all of these cases, however, is the fact that they do not truly constitute examples of attempted leverage. The Soviets were punishing "transgressions" that had already occurred, not trying to influence policy in the targeted countries. To be sure, these countries knew that they were risking the loss of Soviet oil by adopting policies that Moscow opposed. But none felt compelled to shift its policies in light of that threat.

The Soviets have never used the oil weapon against the West. At the time of the first OPEC crisis in 1973, Moscow could have caused short-term disruptions in Western supply by withholding its oil. Instead, the USSR increased its sales on short notice. It was motivated by the incentive of a large short-term profit, and by the possibility of increasing exports to western Europe in the long term. This is one striking example of the Soviets' choosing a long-term economic advantage over a short-term tactical one in their relations with the West.

Moscow is not known to have employed the gas weapon in the past. The abnormally cold winters of 1976/77 and 1978/79 caused a 25 percent reduction in the supply of Soviet gas to the EEC, in part because the Council for Mutual Economic Assistance (CMEA)—eastern Europe's "Common Market"—needed more gas for itself and also because of a pressure drop in the pipelines. These losses did not violate contractual guarantees and were made up that summer. [11] In 1984, during the coal miners' strike in Great Britain, Moscow hinted that it might cut off the flow of gas to the United Kingdom in an expression of "solidarity" with the strikers until the dispute was settled.[12] This threat never became reality, nor did it have any effect on the outcome of the conflict. On the basis of past experience, the Europeans have come to consider the Soviets to be reliable suppliers of energy, and certainly a safer source than the OPEC nations for oil or Algeria for natural gas.

Of course, foreign policy makers cannot go by experience alone in formulating strategy—they must look at the present state of their overall relations with a supplier and take into account diverse scenarios for the future, even if they are remotely theoretical. Some strategists fear that, given the potential to cause short-term disruptions, the Soviets might use gas subtly to pressure European governments for particular concessions on a host of political issues. This pressure could be increased by support from industrial-financial lobbies in the West with important stakes in trade with the Communist bloc.[13] The logic of this argument seems strained. Because Soviet leverage would be only short-term, European leaders would likely ignore any Soviet cutoff threats, knowing they could weather the loss of gas and avoid serious long-term damage. The Soviets, on the other hand, could not be certain how their threat would be received. Not only could the Europeans ignore it, they might become markedly less accommodating to Moscow on a variety of issues such as arms control. A cutoff would certainly jeopardize future energy contracts between the USSR and the West, which is clearly undesirable from the Soviets' perspective, given their need to export energy resources in exchange for Western technology and agricultural products. The use of gas leverage to exert political pressure might well prove far more disadvantageous to the Soviets than to the EEC.

In case of war, however, Moscow would certainly stop the flow of its gas to the West and recovery time by the EEC could be a key strategic concern. Because of dual-firing capability and stockpiling, such a cutoff would not unduly harm the West's defense capacity in a conflict lasting one month or less. Even during a longer confrontation, gas shortages would be unlikely to have serious adverse effects because gas is used primarily in the residential and commercial sectors. The lack of heating in civilian areas during a cutoff, particularly in mid-winter, could provoke something as subtle as a loss of morale among the population. This hardly seems cause for great concern; in wartime, countries adapt to much greater hardships. Only if the USSR succeeded in simultaneously halting the flow of its gas *and* Middle East oil to western Europe might a grave dilemma exist. But the loss of oil alone would do so much damage that a gas embargo would become an almost tangential problem.

In any event, this scenario assumes some sort of conventional war, which most military experts consider extremely unlikely. Even if the Soviets were confident that they could win a war, sustain acceptable losses, and prevent the United States from intervening, they would probably have little desire to take on the burden of occupying western Europe, considering the problems now plaguing them with their own satellites and in Afghanistan. The last scenario is thus a nuclear confrontation. In that case, natural gas, like almost everything else, would become irrelevant.

There is a situation that falls between the use of gas for political leverage and a cutoff in the course of a war: a peacetime crisis. Here, a supply disruption would not be employed as a means to a given end, but rather as a tool of retaliation or escalation that could ultimately lead to war. When the Euro-Siberian gas supply contracts were signed in 1982, some critics of the pipeline in western Europe envisaged hypothetical "doomsday" scenarios. One of these saw the Soviets enter East Germany to crush a popular uprising. In response, the alliance—

miraculously united—boycotts all trade with the USSR. Moscow retaliates by turning off the gas tap.[14] Would a war ensue? The question is almost irrelevant. A Western trade boycott would no doubt include the cessation of gas purchases from the USSR. A Russian foray into East Germany that provoked an allied trade embargo might suffice of its own accord to prompt a military conflict between the two blocs. If war did not result, a gas cutoff, while another step up the ladder of escalation, would at worst provoke a Western countermeasure such as a blockade. Even if East-West gas trade did not exist, the Soviets could use other means of escalating the conflict if they so desired, such as troop mobilization or increased intervention in Third World countries. Criticizing the pipeline for being a potential tool of escalation makes little sense if other such tools are available. These scenarios are far-fetched.

Fears concerning potential Soviet energy leverage over the EEC thus appear to be greatly exaggerated. Prior to the pipeline, some regions in western Europe, such as Bavaria, the Rhineland, and the Saar, were already heavily dependent on Soviet gas. Yet concern over this problem never materialized before construction was begun on the Siberian pipeline. Also, gas from the pipeline will mostly replace the gas western Europe would have received from the Soviet Union had the triangular deal gone ahead. That project was not criticized as a strategic blunder. One wonders how sincere the professed fear of leverage really was, and whether it might not have been used as a tactic to force the cancellation of the project.

In fact, as energy expert Jonathan Stern has argued, the Siberian pipeline will not change the Soviets' position in supplying western Europe with energy. Throughout the 1970s, Moscow exported significant amounts of oil to the EEC. Soviet oil trade is now decreasing, and probably will be increasingly replaced by gas exports over the next decades. The 55 bcm/year of gas that the USSR will sell to the EEC by 1990 is not, on a BTU basis, so very different from the 1 million barrels/day oil equivalent of oil *and* gas that West Germany, France, Italy, Austria, and Finland combined received from the Soviets during the 1970s. The real change will simply be in the composition of exports, with the mix shifting from a preponderance of oil to a preponderance of gas.[15]

Most European experts and government leaders believe that dependence was the source of unjustified fears. "Do people seriously think that we did not carefully calculate our dependence on the Soviets?"

asks Helmut Schmidt, West German chancellor at the time of the pipeline crisis. "The likelihood of a cut-off is much less than that of a disruption in the supply of Middle East energy. If something should stop the flow of Soviet gas, the problems would not be significant. I was always convinced that dependence was an artificial argument brought up by ideologues opposed to the pipeline."[16]

HARD CURRENCY EARNINGS

Soon after gas began flowing through the Siberian pipeline, the *Wall Street Journal* noted that "So frantic has Moscow become to sell its gas that it recently tried to create a spot, or open market for natural gas much like the one that exists for oil." The Soviets have also been selling large quantities of gold. Between 1980 and 1981 alone, sales of gold rose from 90 million to 200 million tons.[17] This reflects not only the USSR's need for hard currency to purchase Western technology and foodstuffs but also indicates that country's inability to compete on the international industrial market. Indeed, so inefficient is Soviet production and so archaic is the country's industrial base that the only way to balance the USSR's import/export ratio is by bleeding the soil of its natural wealth in order to sell those precious raw materials to the world. Critics of the pipeline have argued that the Euro-Siberian is a boon to the Soviets because it will supply them with a substantial and consistent flow of hard currency. This will allow resources to be diverted to the military that otherwise would have been used elsewhere.

As we will see in more detail, the Soviets have not been known to curtail their military expenditures, even in the face of economic stagnation and resulting consumer penury. To assume that, without the pipeline earnings, Moscow would reduce military spending is unfounded. The Soviets would be more likely to tighten the belt on the civilian economy. Be that as it may, gas revenues will ease whatever choice exists between guns and butter. Some savings may indeed be passed along to the military. Moscow needs hard currency for its purchases abroad, and the Soviet economy can only benefit from the sale of gas. So the question remains: Just how important will earnings from the Euro-Siberian be to the Soviets?

Before several EEC countries decided to take smaller quantities of gas or none at all from the Euro-Siberian pipeline, it was estimated that Moscow would earn as much as $15 billion a year from its gas sales to the West. Now that all the contracts have been signed and it is clear the Soviets will provide western Europe with less gas then planned, and at lower prices, these estimates have been substantially revised downward. The Department of Defense calculates Soviet hard currency earnings from the pipeline at a maximum of $7.6 billion a year; other analysts are less optimistic, predicting an annual revenue of $4 to $5 billion.[18] In any event, this level of earnings will not be achieved until the mid-1990s. The pipeline must first reach full capacity, and Moscow must make interest and principal payments on the huge loans taken out in connection with the pipeline. Until the mid-1990s the Euro-Siberian is unlikely to generate more than $2 billion a year in net earnings.[19]

Soviet income from the pipeline must be balanced against old debts now coming to term, as well as channeled into new investments in the energy sector. At least $900 million in Soviet borrowings from the late 1970s and early 1980s matured in 1985, and more debts will have to be serviced or repaid in the coming years. Several large Western financial institutions, including American banks, have discretely helped Moscow to refinance these loans.[20] The Soviets' gas pipeline expansion program, of which the Siberian is just one part, will cost approximately $50 billion from the mid-1980s to the mid-1990s.[21] According to the IEA, profits from the Siberian pipeline will serve mostly to finance this project.

Perhaps more important in the long run is the fact that Moscow's hard currency windfall from the pipeline will be set off by a projected drop in oil exports, all the more so if the price of oil remains low. In 1977, the Central Intelligence Agency (CIA) released the first in a series of controversial studies claiming that an inevitable decline in Soviet oil production would mean the loss of about $13 billion in foreign exchange per year by 1990—some 50 percent of total Soviet export earnings, including those derived from the sale of gas, gold, and arms. These studies were sharply contested, notably by the Defense Intelligence Agency, which predicted continued oil exports into the 1990s.[22]

The CIA prediction, though overly dramatic, is proving to be the more accurate view. Soviet oil production began to decline in 1985, dropping below 12 million barrels a day for the first time since 1980. While the impact on exports has been partially mitigated by a decrease in domestic consumption brought on by the economic recession in the USSR, exports too are suffering now. The USSR cut its sales of oil and oil products to the non-Communist world by almost a quarter in 1985, to roughly 597 million tons.[23]

Moscow is clearly worried. In February 1985, the leadership dismissed the minister of the petroleum industry. The best guess now is that export revenues will decline through the early 1990s, though they will still amount to $6 to $7 billion a year. This estimate must be viewed with some skepticism because of the great number of variables involved, including the price of oil, Soviet domestic consumption, and oil sales to other East bloc countries. The last variable presents a true dilemma for Moscow. The USSR would rather sell oil to the West, because it usually grants its satellites preferential prices, which means less hard currency for Moscow—or none at all when the oil is bartered. Yet the satellites desperately need cheap Soviet oil, and Moscow needs the leverage obtained by keeping its "allies" dependent on it for energy. In 1985, the USSR found a way to keep selling oil to its neighbors and still make a substantial profit. It "convinced" the members of COMECON to pay for Soviet crude at the average of the world price for the previous five years—about $31/barrel. However, most of the satellites will not long be able to afford this deal.

After it pays off its debts, finances other energy projects, and absorbs lost earnings from oil, Moscow will have weak foreign currency reserves. Yet its needs for hard currency are constant and growing.

The U.S.-Soviet five-year grain deal signed in 1983 obliges the USSR to buy 9 million metric tons of American grain each year. It buys about another 35 million metric tons from other sources. This requires between $10 and $20 billion in hard currency.[24] Unless greater efficiency is somehow injected into the Soviet agricultural system or the weather becomes miraculously benign, expenditures of this magnitude for grain will continue into the foreseeable future. Then, there are the tens of billions of dollars that Moscow spends each year in the form of economic and military assistance to its allies. These expenses are also likely to remain constant.

The dilemma is compounded by the USSR's inability to produce and sell quality export goods. Raw materials and arms are the only salable commodities they possess. A rigid, state-run economy in which consumers have no voice and producers no incentive to innovate and manufacture first-class products means that the Soviet Union cannot compete against more flexible economies.[25] Here, too, unless Moscow seriously implements the economic reforms it has hinted at since Yuri Andropov's brief tenure, it will have to continue to sell large quantities of raw materials abroad—and thus deplete its natural resources—in order to earn hard currency.

Given all these facts, assessing the importance of hard currency earned through the sale of gas to the West depends on one's predisposition to view the glass as half empty or half full. Looking at the larger picture, $2 billion/year in pipeline revenues now, and $5 billion by 1995, may not seem abundant. On the other hand, one could argue that with oil revenues dwindling and bills to be paid, every billion dollars counts. Seen in this light, hard currency earnings from the Euro-Siberian are very important.

Either way, the West has little cause for concern. If the hard currency windfall from the Euro-Siberian has indeed been exaggerated, then it is not a strategic problem for the West. If, however, those earnings are essential to Moscow, then what better guarantee could the West have against supply disruptions? Moscow would have to be hard pressed to forgo $5 billion/year in hard currency if it turned off the gas tap. At the same time, the more hard currency the Russians have, the more they will buy from the West, including the United States. Our farmers, our manufacturers, and our lopsided balance of trade can all benefit. This hard-currency-for-gas equation is a classic example of interdependence: if one party attempts to violate the agreement in order to hurt the other, it will do equal damage to itself.

It is true that the distribution of the effects of a disruption in gas supply over time would vary. The loss of gas would have an immediate impact on the EEC but would gradually decrease. The loss of hard currency would affect the Soviets in a much more delayed fashion because of the subsidies and lengthy repayment schedules that were part of the gas deal.[26] Nevertheless, it is difficult to conceive of a case in which the stereotypically cautious Soviets would sacrifice long-term stability for short-term gain—the stakes would have to be extremely high to warrant such a risk. Energy exports represent two-thirds of all the USSR's hard currency earnings, while the supply of Soviet gas to the EEC will make up about 6 percent of the Europeans' total energy needs. Which side would be more seriously affected by blackmail?

THE TRANSFER OF ENERGY TECHNOLOGY

Critics of the Euro-Siberian have argued that the sale to the USSR of equipment for developing Soviet gas fields and building the pipeline was a strategic mistake. They believe that this technology gave the Soviets and their allies access to gas that otherwise would have been extracted at a much higher cost—if, indeed, it could be extracted at all. In turn, they affirm, the savings from producing more energy at a lower cost were passed along to the military. These are legitimate concerns.

It is clear that Western technology, including that sold to the Soviets for the Euro-Siberian, has dramatically improved the USSR's energy infrastructure. The five domestic lines now being constructed have benefited from pipeline technology and Western techniques. This translates into more gas for Moscow and its satellites, and probably more energy for the Red Army. Here again, however, it is important to note that Moscow's need for up-to-date Western technology for the continued development of the energy sector serves as an added guarantee against malicious disruptions in the supply of Soviet gas.

It is also true that there is some direct competition for resources between the energy and military sectors. High quality steel used by the military is lacking in the energy sector, which needs it to manufacture steel pipe. In theory, to get the steel for energy production, Moscow must either take it away from the military or import it.[27] With Moscow's emphasis on military output, however, an embargo on the sale of Western pipe to the USSR might hurt the energy sector but would do little direct damage to the military.

In any event, promoting Soviet energy production is not necessarily undesirable. Increased production enhances global stability by adding to the overall world energy market supplies.[28] Moreover, it seems strategically sound for the West to draw on the energy reserves of other countries before it draws upon its own. As Thierry de Montbrial, director of the French Institute for International Relations, told the Joint Economic Committee of Congress:

> It makes sense to save the Western energy resources or reserves and to draw on the Soviets'. I am surprised that this point is not made more often. But after all, is it not good to keep in reserve the oil we may have in the North Sea, for instance, for later use, and for the time being draw on the Soviets' resources?[29]

Some Soviet officials have expressed reservations about wasting the nation's natural resources on the West. At a conference in 1982, Iulii Bokserman, then vice-chairman of Gosplan's Projects Evaluation Commission, asked other officials, "Are we squandering our gas, or at any rate are we using it sufficiently rationally, by sending it to the capitalist countries?"[30] This comment reflects a concern that the Soviets have had since they first began selling large quantities of raw materials to the West in the early 1970s. In 1972, Anastas Mikoyan, the former Soviet head of state, told East-West trade specialist Samuel Pisar:

> We are ready to sell you our gas and our oil, but not too much. We will have to confront tomorrow the energy problem you are facing today. Our children, our grandchildren, would never forgive us for exhausting the vital resources they will need one day to heat and light Moscow, Leningrad and Kiev. . . .[31]

It is also possible that helping the USSR satisfy its own energy needs will make a Soviet foray into the Persian Gulf less likely.[32] When the pessimistic CIA projections for Soviet energy production were still in vogue, Secretary of Defense Caspar Weinberger seemed to advance that point when he said, "With the Soviets becoming an energy-importing nation in the next few years, the worry is that they would move down through Iran, Iraq and Afghanistan and try to seize the oil fields."[33] The Soviets did of course invade Afghanistan. Whether that initiative was designed to strengthen the USSR's strategic position in the Persian Gulf, or simply to bring back into line a satellite Moscow believed was escaping from its orbit, is the subject of endless debate. To be sure, the USSR might seek to control the Persian Gulf region for purely strategic reasons, and not just to satisfy its energy requirements. But Moscow is likely to behave less recklessly in this part of the world if it does not have to worry about finding energy for itself. Presumably, the Soviets draw up a balance sheet of anticipated benefits and detriments before acting. If they remain energy self-sufficient, oil will occupy a less prominent place in their calculations concerning the Persian Gulf.

Another way to look at the criticism of technology transfer is to assess what would have happened to the pipeline if the Reagan embargo had remained in place or, alternatively, what would have become of Soviet energy production without Western technology. An answer to the first part of this query goes a long way toward answering the second part.

In the short run, the Reagan ban would have caused delays in building the pipeline. Even the partial embargo lasting only a few months certainly hindered the installation of compressor stations. Richard Pipes argues that "The United States had a virtual monopoly on the production of high capacity compressor technology. We were pretty certain we could delay the pipeline's completion for a long while."[34]

There is little doubt, however, that the pipeline would have been completed despite the embargo. As early as August 1982, the CIA had concluded that "the USSR will succeed in meeting its gas delivery commitments to Western Europe" even if the embargo was maintained. The Agency said that the USSR had three options. It could have begun deliveries on schedule in late 1984 by using existing pipelines that had an excess capacity of 6 bcm. Using some combination of Soviet and western European equipment, Moscow could have a working new line by late 1985 that would reach full volume in 1987. Finally, at substantial cost to the domestic economy, the USSR could have diverted construction crews and compressor station equipment from the new domestic pipelines being built to the export pipeline, or devoted one of the domestic pipelines to export use.[35]

Other experts believe that the USSR eventually would have developed equipment comparable with the embargoed technology. The Soviets are traditionally good at producing turbines. Heavy-duty hydroelectric units like those used to operate the great dam at Dniepropetrovsk or the Aswan dam in Egypt, while not state of the art, are durable and highly regarded by Western engineers. Ed Hewett of the Brookings Institution, a leading expert on the Soviet energy industry, has shown that the Reagan embargo pushed the Soviets to develop and produce their own high quality technology. Although they could not compensate for the loss of Western technology with indigenous production while the American sanctions were in effect, they showed signs of bridging the technological gap even in that short period of time. Proptypes of 16 megawatt (Mw) and 25 Mw turbines were developed, and the Soviets now seem capable of mass-producing them at the Nevsky factory in Leningrad, though these units are probably inferior in quality and efficiency to their Western equivalents. In addition, some experts believe that such delays as did occur in building the pipeline were not the result of American sanctions, but rather of Soviet inefficiency.[36]

Thus, even a coordinated embargo would only have forced the Soviets to produce the pipeline technology themselves. This is precisely what happened following the Friendship pipeline embargo in 1962. The Kennedy administration compelled the West Germans to cancel the sale of large diameter steel pipe to the Soviets that was to be used for the construction of an oil pipeline. In response, Moscow increased its production of 40-inch steel pipe, which grew from nothing in 1961 to 600,000 tons in 1965. Meanwhile, Western equipment suppliers lost part of their Soviet market.

Fabio Basagni, an Italian energy expert who was codirector of the Atlantic Institute in Paris, makes a sober judgment of the case against denying the Soviets energy technology.

> These kind [sic] of things [technology embargoes] only bring out the Russian autarkic reflex. Then they do it themselves, though maybe not as well as we could. But the West loses markets. It makes more sense for us to help them and for them to help us than for either side to go it alone.[37]

THE FORCED LABOR QUESTION

One aspect of the Euro-Siberian project was the subject of a surprisingly brief debate—the alleged use of forced labor by the Soviet Union to build the pipeline.

Forced labor has been an all too common feature of the Russian penal system since the days of the tsars. After the Revolution, in view of the tremendous size of many Soviet industrial projects, such a work force was deemed important for the number of workers it provided upon command and at little cost for their efforts. Under Stalin, what had been a cruel practice became monstrous. The forced labor work force is estimated to have expanded from about 30,000 people in 1928 to nearly 14 million in 1953.[38] Many of these workers were purge victims and former prisoners of war whom Stalin did not trust after their "exposure" to the West.

With Khrushchev's de-Stalinization policy, the use of forced labor decreased sharply. Nevertheless, in February 1983, the State Department estimated that there were still 4 million forced laborers toiling in 1,100 camps across the Soviet Union, most of them located near large-scale construction projects.[39] Just how many of these are "political" prisoners and how many are convicts who have been put to work is difficult to say with any certainty, because the Soviets often incarcerate those we would term political prisoners for common crimes. Andrei Sakharov estimated in 1974 that there were between 2,000 and 10,000 political prisoners in the USSR. The CIA has put the number at 10,000, as did Amnesty International in 1975. More recently, a detailed 1984 study by Radio Free Europe, which is not known for pro-Soviet sympathies, provided 863 names of people it said were imprisoned for political crimes, 202 of whom were in psychiatric hospitals.[40]

Amnesty International alleges that minimum health, nutritional, and safety standards are not being met for the majority of prisoner-workers in the Soviet Union, despite the fact that the USSR has signed several international agreements guaranteeing such standards.[41] According to a British study, many forced laborers have been used on pipeline projects in the Ukraine, Kazakhstan, and Siberia, though it did not specifically mention the Euro-Siberian project.[42]

Since Western policy makers were aware of these allegations when the Euro-Siberian was first discussed, they could safely assume that some forced labor battalions would almost certainly be used to build the pipeline. Yet it was not until a little-known Frankfurt-based human rights group issued a report in August 1982, claiming the use of slave labor on the pipeline, that any criticism of this aspect of the project was made. The report referred specifically to past gas projects in Siberia, but also contained personal testimonies about use of forced labor on the Euro-Siberian itself. A few weeks later, American Secretary of Defense Caspar Weinberger declared that "the evidence has been mounting that the Soviet Union may be using slave labor" to build the pipeline. "The evidence is not conclusive," said Weinberger, but it is "profoundly troubling."[43]

Government officials in western Europe felt compelled to respond to these charges and ordered that studies of the problem be undertaken. At the same time, the Soviets invited the International Labor Organization (ILO) to visit pipeline construction sites. The minor outcry in the West subsided in anticipation of Western government and ILO findings. However, Moscow denied the ILO total freedom of access to the pipeline, and the ILO refused to undertake the study in the absence of such a guarantee. In September 1983, the Soviets announced that all the pipe had been laid for the project, and that there was, consequently, no reason to look for forced labor.[44] The brutal logic of this argument did not escape the ILO, which abandoned its mission. As for the individual government reports, most of them were canceled or given low priority pending completion of the ILO study. Some embassies did report back to their governments. The French mission in Moscow told the Foreign Ministry in Paris that the use of forced labor on the pipeline was "probable, but difficult to prove."[45]

Thus, little was done in the West about a truly dark side of the Euro-Siberian project. It is perhaps not very surprising that the western European governments chose to ignore what evidence was presented to them concerning forced labor. This problem had never hindered East-West trade before, and the Siberian pipeline was too important to them to be sacrificed for moral considerations. Why the

United States government made little fuss about the issue while it made so much over other issues involved is more perplexing.

Assuming that some forced labor was used on the Euro-Siberian, it is imperative to inquire whether the West should have continued its participation in the project. Clearly, the free world could have avoided the taint of benefiting from forced labor by having nothing to do with the pipeline's construction or its future gas flow.

Some would invoke "raison d'état" to ignore this dilemma. The pipeline, after all, bolstered Western economies and promoted energy diversification; sometimes it is necessary to disregard one's principles. If the West stopped trading with all the countries that abuse the rights of their citizens, it would deprive itself of myriad necessary resources and goods, ranging from oil and gold to magnesium and uranium.

Moreover, this argument continues, governments do not often concern themselves with the fact that other governments deprive their own people of human rights, unless the country in question is an adversary. China, for example, which allegedly employs many more forced laborers than the USSR, is not criticized for this policy, seemingly as a result of its improved relations with the West. Numerous Asian, African, and Latin American countries with which the West trades exploit and underpay their people, inflict hazardous working conditions upon them, and permit child labor. It would be hypocritical to take a stand on the forced labor issue. At any rate, man's inhumanity to man is common knowledge. Such is the human condition.

This view, while admittedly grounded in a realistic perception of the way nations tend to act and interact, is not particularly satisfying. If we are truly concerned about the lot of slave laborers in the Soviet Union, the pertinent question to ask is whether renouncing the project would have improved their condition. On the basis of past experience, it seems clear that polemics, heated rhetoric, and economic sanctions have had little success in forcing Moscow to respect the human rights of the Soviet citizen; on the contrary, such actions tend to be exercises in futility. The Soviet Union cannot be compelled by force or confrontation to modify its behavior to meet our desires.

That the Euro-Siberian may have been built with the muscle of forced labor is a tragedy. But is is possible that increasing economic ties between East and West may one day loosen the grip of Soviet totalitarianism upon its people. Seen in this perspective, the Euro-Siberian has definite short-term costs, but the ethical payoff in the long term may render them acceptable, if not less grievous. The pipeline poses a moral problem with no clear-cut answer. It may prove to be in vain, or naive, to do so, but one must try to keep alive the hope that the benefits to East and West that will flow from the Euro-Siberian will help relax tensions between the two blocs and perhaps help curb brutal Soviet domestic policies.

The role of trade with the Soviet Union and its satellites has caused an ever widening crack in the Western alliance since its infancy. The most powerful member of the partnership, the United States, has typically tried to cripple the Soviet Union, or at least contain its economic and military growth, through trade restrictions. Western Europe, although resolutely tied to America by common ideological, social, political, cultural, and economic bonds, is wary of its precarious geographical position between the two hostile superpowers. For this reason, and in order to take advantage of their natural economic complementarity with the East, the Europeans have sought to build lasting commercial ties between themselves and the Soviet bloc. This divergence in inter-alliance policies has been brewing for decades. The Siberian pipeline project finally forced it to boil over.

AN ALLIANCE DIVIDED

5

Seeds of Discord

Trade with Russia has been an issue in the West since well before the 1917 Revolution.[1] When the Bolsheviks took power, the debate simply intensified. In 1920 and 1921, British Prime Minister David Lloyd George and Deputy Chairman of the Soviet Central Government L. B. Kamenev set the parameters for an argument that continues to this day. Lloyd George, a proponent of increased trade between East and West, believed that, having failed

> . . . to restore Russia to sanity by force . . . we can save her by trade. . . . Trade, in my opinion, will bring an end to the ferocity, the rapine and the crudity of Bolshevism surer than any other method.[2]

Kamenev was also a supporter of greater commercial ties between his nation and the West, but for reasons diametrically opposed to those of Lloyd George.

> We are convinced that the foreign capitalists who will be obliged to work on the terms we offer them will dig their own grave. Foreign capital will fulfill the role Marx predicted for it. . . . With every additional shovel of coal, with every additional load of oil that we in Russia obtain through the help of foreign technique, capital will be digging its own grave.[3]

In the early 1920s, trade between East and West was modest. Lenin's New Economic Policy sought to attract foreign investment. But mutual suspicion restricted trade, as did a lack of Soviet exports to pay for Western imports. Toward the end of the decade, trade increased as Stalin's First Five-Year Plan (1928–32) called for significant imports of Western technology to further Soviet industrialization. Some of the Soviet Union's largest industrial complexes were built during this period with Western help, mostly German and American. Packard, Ford, General Electric, International Harvester, and U.S. Steel were among the major American companies to undertake projects in the USSR. By 1931, the Soviets were absorbing two-thirds of all U.S. exports of agricultural equipment and power-driven metalworking equipment.[4]

The Great Depression curtailed the further expansion of East-West trade. The West's economic difficulties bolstered Stalin's view that increased commercial ties with the capitalist countries could have a destabilizing economic impact on the Soviet Union. Equally important, the Soviet leader feared integrating his country into the world economy; such a development, he believed, might disrupt the USSR's political and social systems.[5]

World War II and America's lend-lease program to support its allies in the fight against Germany promoted a dramatic upswing in trade between the United States and the Soviet Union. Under lend-lease, such trade amounted to several billion dollars every year.[6] But the Allies' victory in 1945 brought an end to this partnership of necessity between the United States and the USSR, and a return to the uneasy prewar relationship. Immediately following the war, it was difficult to foresee the extent to which that unease would grow. But a new balance of power in the world had been created, and gradually it crystallized into the Soviet and American spheres of influence. This balance and the relationship among the countries in each camp have since dictated postwar trade relations between East and West.

*

THE COLD WAR YEARS (1945–62)

The roots of the Cold War are diverse and still the subject of substantial debate. In essence, the establishment of antagonistic relations after World War II can be attributed to the underlying differences between the Eastern and the Western economic, political, and social systems, leading to the view of policy makers on both sides that the survival of one system necessitated the death of the other.[7] During the 1920s and 1930s, internal western European problems diverted attention from East-West tensions. With the defeat of Germany and the emergence of the United States as the dominant power in the West, the competing systems found themselves face to face.

Despite the agreement at Yalta between Franklin Roosevelt and Stalin to "divide" the world into Eastern and Western spheres of influence, a series of crises developed between the United States and the Soviet Union as early as 1946. The two powers confronted each other over Manchuria, Iran, and Turkey. Disputes then arose in China, Korea, and eastern Europe.[8] Each side quickly attempted to consolidate its sphere of power. The East-West division became a fait accompli with the creation of the North Atlantic Treaty Organization (NATO) in 1949 followed by the establishment of the Warsaw Pact in 1955.[9]

The East-West split naturally affected trade relations between the two blocs. At its core, U.S. policy sought to hinder the development of the Communist economies by denying them the benefits of trade. A form of economic warfare, though not acknowledged as such, became U.S. policy. Following Washington's lead, the alliance countries created complex national and multilateral networks of trade restrictions in the form of import and export controls. From the outset this U.S. policy was in direct conflict with the western European view. A U.S. government study of East-West trade relations noted:

> Few Western European or Japanese statesmen or businessmen shared the underlying assumption, or for that matter, the ultimate objective of the embargo. . . . Allied differences with the United States rested both on policy and economic interest. Europeans could simply not accept the view that denying trade would put an end to Communism or even curtail Communist countries' development. In more pragmatic terms, trade with Eastern Europe was a matter of no small consequence to our Western European partners.[10]

Prior to 1945, western Europe had proceeded to develop strong and consistent ties with the East. Between the two world wars, roughly 75 percent of the trade of eastern European countries was with western Europe, whereas trade within the central and eastern part of the continent, which was later to become communist, accounted for only 15 percent. Immediately following World War II, the western Europeans sought to multiply their trade links with the East. In 1947, Great Britain, and in 1948, Italy, signed trade agreements with the Soviet Union; in 1949, both countries began to purchase Soviet oil.[11]

By 1950, however, East-West trade had been dramatically curtailed. One reason was Stalinist xenophobia. Another was an overt policy of autarky. On paper, Stalin's strategic vision was compelling.

The USSR would create an independent trading zone by removing eastern Europe, with its 350 million consumers, and ultimately Communist China, with its 800 million consumers, from the world economy. In this manner, the Soviet Union could strengthen its hold over the satellites and, Stalin believed, strangle the capitalist countries by amputating their markets. In 1949, at Moscow's initiative, the East bloc countries except Yugoslavia formed their own common market, the CMEA. One year later, more than 60 percent of the eastern European nations' total trade was within the Communist bloc. By 1953, that figure stood at 80 percent.[12] However, it quickly became clear to the Communist leadership that the East bloc's economic backwardness in relation to the West made autarky impractical and counterproductive. Following Stalin's death, the Soviet Union under Krushchev stopped trying to reinvent the wheel and sought to reenter the world market to some extent.

While the Soviets changed direction, Washington continued its quest for an alliance-wide embargo on trade with the Communist bloc. America's dominant position in the newly born NATO forced the European partners down a path they would have preferred not to have taken. NATO, the framework of the Western alliance, was created in order not only to provide a common defense network but also to solidify the Western sphere of influence born at Yalta. In the postwar period, the devastated nations of Europe were incapable of taking any lead in promoting collective security and defense. The dominant policy-making role fell to the United States. America's preeminence continued through the first postwar decades as the major European powers lost their empires, and the wealth that went along with them, through decolonization.

Despite the economic incentive to trade with the East, the Europeans' stake in following Washington's lead was great. The allies had shared values and similar economic, political, and social systems that served as a natural glue for Western solidarity. Partly to ensure the strength of that glue, Washington in 1947 proposed the Marshall Plan, which helped to rebuild the western European nations into prosperous, liberal economies and sturdy democracies through a massive infusion of economic aid.[13] American legislators immediately sought to condition the continuation of Marshall Plan assistance on the allies' acquiesence to America's East-West trade policy. Western Europe, in economic disarray and dependent upon American aid for reconstruction, was in no position during the first postwar decade to exert any meaningful influence on the formulation of trade policy toward the East.

The American legislative initiatives in question were the Economic Cooperation Act of 1948 and the Mutual Defense Assistance Control Act of 1951. The former required recipients of Marshall Plan aid to assure the United States that they would not export items that Washington considered strategic to eastern Europe. The latter, known as the Battle Act, was more stringent still. It threatened countries that knowingly exported arms, implements of war, atomic energy materials, and other strategic goods to the East with a cutoff of American military and economic assistance, including aid not tied to the Marshall Plan.[14] The Battle Act sanctions were never applied, and although the allies did sell the East certain goods that Washington viewed as strategic, the bill on the whole had a deterrent effect on western European-East bloc trade.[15]

The United States did not attempt to enforce export controls solely by threats. In late 1949, the NATO countries, except Iceland and with the addition of Japan, agreed to establish a multilateral group that would oversee export restrictions at an alliance level. The procedures to which this initiative gave rise are in operation today. The so-called Consultative Group Coordinating Committee (COCOM) began to meet at Paris in 1950. COCOM is an informal entity: the group has no charter, is based on no treaty, and its decisions are only "morally," rather than legally, binding. Its task is to establish lists of strategic items subject to embargo or monitoring, and to develop means to assure compliance. Member countries bring a product before COCOM for scrutiny. Experts representing each member decide whether the product should be embargoed, which is subject to unanimous agreement. Yet because the COCOM list is a nonbinding recommendation, it only gains legal force if a member decides to carry out the embargo through its national export control program governed by its own laws and regulations. Once a product is on the embargo list, a country may request that an exception be made so that it can export the controlled item to a specified end user. A unanimous vote is, in theory, required. But here too, because COCOM has no binding legal effect, a country may choose to disregard the denial of an exception request and sell the product in question. The lists are revised periodically and are kept secret—however, certain national export control lists, such as Great Britain's, are known to be nearly identical with COCOM's.[16] Others vary. The American list is always considerably more restrictive than its COCOM counterpart, just as U.S. attitudes in the sharply divided COCOM discussions are far more stringent. The list shrinks and swells depending on the broad East-West political climate, becoming more liberal in periods of détente and more restrictive in periods of confrontation.

The allies also developed restraints on trade at the national level. In the United States, Congress imposed extensive controls on exports, imports, and credits aimed primarily at Communist countries. The Export Control Act of 1949 required that American companies obtain licenses from the government in order to sell their products to communist countries and gave the president authority in peacetime to ban the sale of strategic goods and technology to foreign nations. The Act, subject to periodic congressional review, could be invoked for three basic reasons: (1) to safeguard national security, (2) to promote foreign policy objectives, and (3) to prevent the drain from the United States of materials in short supply. Under the Act, nothing of potential military or strategic value could be exported to the East bloc without a special license from the Department of Commerce. There were almost 1,000 proscribed items by the mid-1950s.[17]

The Trade Agreements Extension Act of 1951 withdrew most-favored nation (MFN) status from all Communist countries except Yugoslavia. This meant that exports to those countries were subject to the prohibitive tariffs set by the Smoot-Hawley Tariff Act of 1930. Imports from the East, already paltry, lost any competitiveness they had had in the American market. Countries entitled to MFN status benefited further from the sharply reduced tariffs subsequently negotiated by the United States in the context of the General Agreement on Tariffs and Trade (GATT).

The Johnson Debt Default Act of 1934 became an instrument in the reduction of loans and credits available to East bloc countries. This law, which bans loans to foreign governments in default on debts owed to the United States, was not originally aimed at the East bloc, but was in practice applied to those countries by virtue of their unpaid war debts to Washington.[18] The entire complex of legislative instruments and governmental policies conceived in the late 1940s and early 1950s established an environment for the checkered course of East-West trade over the following three decades—culminating in the Siberian pipeline dispute.

Despite their desire to expand commerce with the East, western European countries maintained various trade controls. In practice, these were often relaxed to permit bilateral deals. And whereas Washington used its controls openly as a part of its trade embargo strategy against the East, west European legislation existed ostensibly to correct balance of payments deficits.[19] This fact points to the essential difference between the aims of American and western European trade policies: American policy has been based on what can broadly be termed strategic considerations, while western Europe's has been principally concerned with economic criteria. There have been several notable exceptions since the war to which this general strategic-economic dichotomy has not applied. The difference between western European and American motivations in trade with the East, however, merits further elaboration because it illuminates the differences in allied attitudes and perceptions that led to the pipeline crisis.

Unlike the United States, for which foreign trade comprises a small portion of gross national product (GNP), the EEC countries have historically been among the most active traders in the world. Using the English "workshop to the world" model, the Europeans have fashioned economies that work on comparative advantage. This typically means exporting a surplus of finished industrial goods and importing the raw materials and agricultural produce that western Europe lacks.

Following World War II, the United States forced Germany and Japan to redirect their energies into industry and trade, thus compelling them to become trading states. By the 1950s, imports and exports constituted a high proportion of the aggregate national product of most of the countries that were later to form the EEC. Today, exports account for only 8 percent of U.S. GNP, compared to 24–31 percent for France, Italy, Britain, and West Germany.[20]

Given an economic profile in which foreign trade is the outstanding feature, it is not surprising that western Europe has sought to establish strong commercial ties with the East. The fact that the nations of eastern and western Europe share a common landmass and ancient cultural ties is a further stimulant to trade. The two parts of Europe make a particularly natural economic "fit." The East, disadvantaged by a tardy industrial revolution, has sought to accelerate its economic growth. For the West, eastern Europe has constituted a new market for finished products and industrial goods, as well as a supplier of energy, minerals, and food. Despite the sharply divergent systems of government and management, the eastern and western European economies are too complementary to ignore the advantages that commerce can bring to both sides.

It would be simplistic to view western Europe as a monolithic camp with a common commercial diplomacy. Great Britain, France, and Italy, on the one hand, tended to separate economics and strategic considerations, viewing trade with the East as the province of the market and not that of foreign policy. West Germany, on the other hand, placed emphasis on the political aspect of trade, using it as a tool for conducting its overall relationship with the East, as did the United States. But unlike Washington's preponderantly negative approach, Bonn saw in commerce with the East a positive means to strengthen ties to West Berlin, East Germany, and the whole of eastern Europe. Under Chancellor Konrad Adenauer in the 1950s, trade links were created with the East and then used to wring political concessions from Moscow in the recurrent tensions over Berlin and the "other" Germany. This was an early form of the famous linkage theory that Henry Kissinger revived two decades later. The results were mixed. As Angela Stent shows in her study of Bonn's *Ostpolitik*, German linkage attained few concrete goals.[21] Yet the two Germanys moved much closer to each other over the following decades, and West Germany developed relatively strong and lasting ties with eastern Europe.

Economic complementarity between the United States and the Eastern countries was markedly less obvious until the 1970s. America could have become a prime supplier of modern machinery to eastern Europe, as it was for a time before World War II. But the reciprocal relationship that attracted the two sides of the continent to one another was lacking. Whereas western Europe had a demand for Eastern exports, particularly energy, the United States was largely self-sufficient and had little demand for East bloc products. The East had difficulty financing imports from America without substantial credits to purchase U.S. goods and favorable terms of trade to export their own goods. But because the United States had little need for Soviet exports, political concerns dominated the formulation of American trade policy—the extension of credits was the exception and trade barriers were the rule following World War II.

American companies whose products were in high demand in the East could have pressured Washington to facilitate trade, but government opposition was not the only disincentive they faced. The precarious financial status of most Soviet bloc countries usually prevented them from offering cash payment for imported goods. The antagonistic relationship between East and West gave businessmen cause for concern about the safety of any production, maintenance, or distribution facilities they might establish in the East. The long-term stability of the Eastern market also seemed highly uncertain, and consumers at home, caught up in the anti-Communist mood of the 1950s, often boycotted American companies doing business with the "enemy."[22]

The 1950s saw little change in East-West commercial relations, though a modest decrease in overall tension from 1953 on did improve the climate for trade. Stalin's death and the end of the Korean war in that year rendered commercial initiatives easier. In 1954, the COCOM embargo list was reduced from 285 to 170 items at the insistence of the western Europeans, whose economic recovery and decreasing dependence on Marshall Plan aid gave them a stronger voice in shaping Western trade policy.[23] In 1955, West Germany established diplomatic relations with the Soviet Union, laying the groundwork for a 1958 agreement between the two countries to increase trade. Between 1958 and 1961, Soviet-German commerce grew from $163 million to $401 million,[24] making the Federal Republic the USSR's principal trading partner.

Meanwhile, the Soviets began to show an increasing appetite for trade with the West. Nikita Khrushchev and other Kremlin leaders recognized the drawbacks of Stalin's quest for autarky. Despite the boast that the USSR would "bury" the West economically, Khrushchev was not blind to economic realities. Soviet productivity was declining and the impressive growth rates of the early 1950s had slowed perceptibly. Clearly, Western technology could stimulate Eastern economic development and remove some of the burden of the satellite countries from Moscow's shoulders. Hence Khrushchev's call for a dramatic "chemicalization" program that sought to increase the production of fertilizers for agriculture in 1958. This was accomplished with significant help from the West. Between 1958 and 1963, 50 complete chemical plants were delivered to the USSR,[25] primarily from western Europe.

The decade ended, however, with several events that reemphasized East-West antagonism and diverted attention from trade issues. Most noteworthy were the two Berlin crises (1958–60 and 1961–62) and the construction of the Berlin Wall in August 1961.[26] Despite internal disagreements over trade policy, the simmering difference between western European and American attitudes toward expanded trade was held in check, and the West remained fairly united in its confrontation with the East.

A DECADE OF AMBIGUITY (1962–72)

East-West tension reached its peak with the Cuban missile crisis in

October 1962, then eased substantially with the peaceful settlement of the face-off between Washington and Moscow. This development had a paradoxical effect upon the leaders of the alliance in western Europe. On the one hand, the fear of conflict between the United States and the USSR appeared to them to have decreased. The relative relaxation of tensions seemed to bode well for new trade ties. On the other hand, western Europe's particular vulnerability to the antagonistic whims of the superpowers appeared to have increased. Moscow and Washington might not destroy each other, but they could find western Europe a convenient battleground for confrontation, perhaps even for a "limited" nuclear exchange.[27]

Because of western Europe's growing economic power, as evidenced by the end of Marshall Plan aid and the creation of the EEC in 1959, the allies were able to expand trade ties with the East. Given their uncomfortable position between two hostile, if stable, superpowers, the political necessity for such expansion and the diplomatic détente it might bring in its wake increased dramatically. Western Europe, its leaders believed, could serve as a bridge between Washington and Moscow, thus assuring its own security and reasserting leadership in the concert of nations. At the same time, the EEC's secondary role in the East-West geopolitical drama—that is to say, its "regional" as opposed to the United States' "global" responsibilities—gave it additional justification and flexibility for pursuing a pro-trade policy.

Although economic incentive remained the primary motivation, the western Europeans began to use trade more as a political tool than they had in the past. This was especially true for the French. President Charles de Gaulle envisioned western Europe as a third superpower, independent of both the United States and the Soviet Union. But because France was part of the Atlantic system, charting an independent course meant breaking away from Washington's grasp while remaining resolutely in the Western camp. De Gaulle paid a state visit to Moscow in 1966; three years later, French exports to the Soviet Union were up by 70 percent.[28] The French head of state also traveled to Poland in an effort to improve relations with the rest of eastern Europe. He hoped to wean the satellites away from Moscow by making them less dependent on the USSR for trade. The French leader also espoused the view that commerce between the two blocs could lead to greater stability on the continent. "What we need most of all for peace," he said, "is understanding between peoples. People are what count, not regimes. Regimes ultimately disappear."[29]

Throughout western Europe, the mid-1960s witnessed a relaxation of most controls on trade with the East. The British (1964), the French (1966), and the Italians (1967) liberalized their existing import quotas on Eastern goods. They also furnished considerable credit with extended repayment schedules to the East and signed several bilateral trade agreements. The British, for example, granted a 15-year, £100 million loan to the Soviets for the construction of a polyester plant. The Italians extended a $350 million credit over 14 years for the construction of the Togliatti automobile plant in the Soviet Union. And a Franco-Soviet commercial protocol gave Moscow credit with an eight-year repayment schedule.[30]

The West Germans pursued a more ambiguous trade policy. Bonn's attempts at linkage had borne little immediate fruit and tensions with the Soviet Union and East Germany remained high, souring to some extent West German enthusiasm for trade. At the same time, the highly business oriented administration of Ludwig Erhard, the man most responsible for the German "economic miracle," stressed the crucial importance of trade for the Federal Republic's prosperity. Bonn therefore opted for a middle course by downplaying economic ties with the Soviet Union—save for the sale of considerable amounts of oil pipeline equipment—while emphasizing trade with the other East bloc countries. The West Germans opened trade missions in Poland, Romania, and Hungary in 1963.[31] In a broad sense, the mid-1960s witnessed an upswing in commerce between western Europe and the East bloc.

If European trade policy toward the East appeared to be consistently progressing on an ascending slope, the same could not be said for Washington. American policy seemed tied to a political roller coaster. In 1962, Congress extended the Export Control Act to embargo goods that were economically, and not simply militarily or strategically, beneficial to the East. The Act could be interpreted as calling for total trade embargo. Certainly, it was in clear conflict with the prevailing European views both within COCOM and at national levels, which aimed at controls limited only to items of direct military significance.[32]

In November 1962, immediately following the Cuban missile crisis, the Kennedy administration imposed an embargo on the sale of large diameter steel pipe being exported by West Germany and Great Britain to the Soviet Union for the construction of an oil pipeline. It is worth deviating briefly from the overview of allied trade relations with the East to examine this case, for it bears striking similarities to the Siberian pipeline crisis two decades later.

The pipeline in question was to link the USSR's Baku fields in southern Georgia with the principal industrial centers of the satellite countries. The Soviets were experiencing difficulty in producing sufficient quantities of high-grade steel pipe, and therefore contracted with the western Europeans, notably West Germany, in order to avoid construction delays. Mannesmann, already in the front lines of energy-related transactions as the leading German pipe manufacturer, agreed to sell the Soviets 165,000 tons of steel pipe valued at $28 million.

Washington feared that the Friendship pipeline, as the project was known, would be used mostly to supply fuel to Red Army units stationed in eastern Europe, thus greatly increasing their mobility. At a NATO meeting in December 1962, the United States pushed through a majority vote requiring the allies to institute a special licensing procedure for the export of steel pipe to the East bloc. At the same time, Washington pressured Bonn into canceling the pipeline contracts signed by German companies, citing legislation voted by the Bundestag in 1961 that allowed the government to restrict trade for reasons of national security. Despite sharp disagreement within the Bonn government and strong protests from the opposition and the business community, the Bundestag voted to comply with Washington's demand.

As a result, the West Germans lost tens of millions of dollars in equipment orders—Mannesmann alone was obliged to forego $25 million in sales. When the embargo was finally lifted in 1963, the Germans had difficulty recovering their share of the Soviet market. Chancellor Adenauer's decision to submit to Washington's will caused him considerable political embarrassment when British, Swedish, and Japanese companies stepped in to replace German suppliers. The embargo resulted in a one-year delay in building the pipeline. This prompted Moscow to further the development of its own steel pipe industry; whereas the USSR produced no 40-inch steel pipe in 1961, it was producing 600,000 tons a year by 1965, albeit of lesser quality than pipe manufactured in the West.[33]

Despite the Europeans' anger at Washington's attempt to dictate their trade policies, the embargo did not create the same polarization between the United States and its allies that the Euro-Siberian crisis was to cause 20 years later. While the British, French, and Italians vehemently protested the American initiative and refused to go along,

Bonn in the end meekly acquiesced to Washington's demands. This reflected West Germany's continued dependence upon the United States to guarantee its security, and the sentiment that the ugly stain of World War II still made the German state a less than equal partner in the alliance. Ultimately, the embargo created harmful political problems for Adenauer and proved economically more costly to western Europe than to the Soviet Union.

The 1962 extension of the U.S. Export Control Act and the pipeline embargo were significantly counterbalanced by other policy initiatives in Washington that helped to improve prospects for trade. One result of the Cuban missile crisis had been the quick agreement between Moscow and Washington on a nuclear test ban treaty, signed in August 1963. The treaty, by easing U.S.-Soviet tensions, gave President Kennedy leeway to further commercial relations.

In October 1962, amid much controversy, Kennedy announced the sale of 65 million bushels of surplus American grain to the Soviet Union and proposed to extend credit through the Export/Import Bank. The Soviets had experienced a particularly bad harvest that year, and had tried to alleviate the problem by purchasing Canadian grain. The Canadians, while happy to sell, could not provide the quantities Moscow needed. Washington, burdened with an overabundant crop that was depressing domestic prices and imposing high storage costs, agreed to make up the difference.[34] Taken together, the grain sale and the Friendship pipeline embargo ironically foreshadowed the salient features of the Siberian crisis. First, the United States pressured its allies to cut energy trade with the Soviet Union in the name of security. Then, Washington turned around and sold the Soviets wheat in order to please American farmers.

The wheat deal was a turning point for America's trade policy toward the East: commerce suddenly began to be viewed as a means for relaxing tensions between the superpowers, as well as a potentially profitable undertaking. Businessmen saw the sale as an implicit presidential seal of approval for increased trade with the Communist world. This was a welcome development for the business community, which had become frustrated at the idea of losing a potentially enormous market to the western Europeans.[35] Paradoxically, it also signified an American reconciliation with European views on the issue.

President Johnson continued the policy embarked upon by Kennedy. In December 1964, he gave a highly controversial speech expressing his belief that America should seek to "build bridges to Eastern Europe—

bridges of ideas, education, culture, trade, technical cooperation and mutual understanding for world peace and prosperity.''[36] He supported his lofty rhetoric by creating a special committee to investigate areas for possible East-West economic accords.

Certain members of Congress followed the president's lead. In 1966, Senate majority leader Michael Mansfield, along with Sen. Warren Magnuson and Sen. Jacob Javits, introduced a bill called the East-West Trade Relations Act that sought to remove legislative constraints on East-West commerce. In a letter to Congress in support of the bill, Secretary of State Dean Rusk wrote:

> The growth in trade between East and West would make the Soviet Union and Eastern Europe increasingly conscious of their stake in stability and improving peaceful relations with the outside world.[37]

Strong congressional opposition, however, killed not only the bill but also subsequent efforts to increase trade with the East. In 1967, the creation of a House task force on trade led by Melvin Laird and Charles Goodell, prompted the Republican leaders in Congress, Sen. Everett Dirksen and Rep. Gerald Ford, to declare that they would oppose increased trade as long as the Soviets continued aiding the North Vietnamese.[38] These developments underscored the rift that had opened between the relatively pro-trade executive and a much more cautious legislature. With one notable exception—the liberal Export Administration Act of 1969—that division was to last through the détente years of the Nixon administration. During the late 1960s, congressional brakes on trade reflected a renewed animosity toward the USSR engendered by the Soviet involvement in Vietnam and the invasion of Czechoslovakia. In 1968, Congress approved a bill banning the extension of credit to countries providing assistance to ''any nation engaged in armed conflict with the United States,''[39] a reference to North Vietnam. This effectively served to prohibit Communist countries from receiving Export/Import Bank credit.

In retrospect, the Kennedy and Johnson years produced ambiguous results in the domain of trade policy. Congress passed no significant pro-trade legislation. Still, both presidents helped to change the psychological climate in favor of greater commercial ties, and even prompted discussions in the conservative Congress with a view to dismantling some of the controls on trade with Communist countries.

In view of congressional unwillingness to support increased commercial ties with the East, it is ironic that the first substantial step in favor of trade during the Nixon administration was taken not by the White House but by the legislature. The Export Administration Act of 1969, which replaced the Export Control Act, eliminated the requirement that the economic potential of a given product be taken into account in the issuance of an export license. It also called for a reexamination of the restricted commodities list. The bill was as important for its symbolism as it was for the concrete measures it mandated: much of the trade warfare language of the earlier act was eliminated, from the title—"Export Administration" became the euphemism for "Export Control"—on down.[40]

Still, Congress's apparent reversal was not clear-cut. The passage of the bill came only after lengthy and harsh debate, notably in the House. In fact, the House simply called for an extension of the 1949 Export Control Act. The eventual compromise resulting in the passage of the new act favored those who wanted to expand trade, particularly a group of liberal congressmen eager for the economic benefits of commerce at a time when the American economy had slowed down appreciably. But congressional approval was nevertheless far from unanimous, and is hardly representative of the legislature's usual position on East-West trade.[41]

Across the Atlantic, the election of West German Chancellor Willy Brandt gave a boost to the proponents of East-West trade. The new chancellor held a strikingly different view of relations with the East than his predecessors. He recognized the failure of previous attempts at linkage. No less committed to German reunification than past West German leaders, Brandt took a more long-term approach and reasoned that such a development would depend on the establishment of a more positive relationship between the Federal Republic and the Soviet Union. Brandt implicitly recognized Europe's postwar boundaries and the separation of Germany into Western and Eastern halves. In return, his *Ostpolitik*, conceived and elaborated by fellow Social Democrat Egon Bahr, brought about improved relations with the Soviets and increased contacts with East Germany. As one scholar has argued, Moscow and Bonn made opposite bets: "Whereas Moscow wanted a ratification of the status quo in order to make it more permanent, Bonn agreed to accept the status quo in order ultimately to change it."[42] The effect of the new *Ostpolitik* was to depoliticize trade in the short term. Economic ties with the East were important in their own right, and would not be used as a linkage tool. In the long run, trade might improve the East-West political climate.[43] Meanwhile, German industry should take advantage of the Eastern market.

By the start of the 1970s, the nations of western Europe were roughly united in their attitudes toward trade with the East, with the Germans in the forefront of the pro-trade movement. Policy in Washington continued to move simultaneously in opposing directions, with the more liberal Export Administration Act being blunted by the Nixon administration's initial opposition to trading with the Soviets. By 1969, west European-Soviet trade was valued at nearly $2.5 billion, while U.S. trade with the Russians was less than $200 million.[44] But in 1972, Nixon became the first American president to visit Moscow, and a new era in East-West trade began.

FROM DÉTENTE TO DISARRAY (1972–82)

The reasons for President Nixon's dramatic trip to Moscow in mid-1972 were many. His overture to the East had actually begun one year earlier with another first—a state visit to China. This initiative had given the president, then plagued by the ongoing war in Vietnam and a shaky economy, a more positive note to sound in an election year. It also made it easier to take a step in Moscow's direction, especially in light of the public's positive reaction to the China trip. By easing tensions with the Soviets, Nixon hoped to reduce military spending, give the American economy the benefit of virtually untapped markets in the East, and, most of all, induce the Soviets to pressure their North Vietnamese clients into seeking an accommodation with Washington.[45]

Trade became a pillar of the administration's détente policy. This made good economic sense. The administration hoped to reverse a negative balance of trade. In 1971, for the first time in almost a century, the United States imported more goods than it exported. The following year, the negative balance reached $7 billion.[46] Some 350 million potential customers in the Soviet bloc and another 800 million in China could do much to alleviate the problem.

Nixon's policy owed much to the strategic vision of Henry Kissinger, his national security adviser who subsequently became secretary of state. The new emphasis on trade was mostly due to Kissinger's conception of commerce as a strategic tool. As he saw it, "By acquiring a stake in the network of relationships with the West, the Soviet Union may become more conscious of what it would lose by a return to confrontation."[47] Kissinger hoped to use linkage as a means of influencing the conduct of Soviet foreign policy. The promise of expanded trade could, in theory, induce acceptable Soviet behavior on many sensitive issues and in many global hot spots. The implicit threat of curbing that trade could discourage Soviet transgressions. Kissinger's aim, however, was to place commercial ties after political gains; he insisted on the "White House's determination to have trade follow political progress and not precede it," thereby ensuring that "economic relations depend on some demonstrated progress on matters of foreign political importance to the United States."[48]

During his Moscow trip in May 1972, Nixon and the Soviet leadership agreed, among other things, to establish a joint U.S.-Soviet commercial commission whose task would be to study areas for cooperation between the two superpowers and to negotiate a trade agreement.[49] By October, negotiations had been concluded. Moscow consented to settle its long outstanding lend-lease debt of $722 million, thus satisfying the Johnson Act requirements and allowing the USSR to receive private bank credits. In return, and subject to congressional approval, the United States promised to give the USSR MFN status and extend normal Export/Import Bank loans to Moscow.[50] The first such credit was granted in February 1973. Over the following 15 months, Washington provided more than $460 million in credit, which financed $1 billion in U.S. exports.

Despite the administration's view of trade as a foreign policy tool, both Washington and Moscow took pains to spotlight the newfound American-Soviet economic complementarity. During the Nixon visit, for example, Soviet President Nikolai Podgorny proclaimed:

> The Soviet Union and the United States are powers that are most advanced in science and technology, they have vast economic potential and rich natural resources. . . . All this serves as a solid foundation . . . to establish Soviet-U.S. cooperation in the most varied fields, to implement large scale projects worthy of the level which the Soviet Union and the United States have reached in the world today.[51]

Following the administration's lead, American companies took a much greater interest in trade with the East, notably in connection with large-scale industrial projects. One area of special interest proved to be the development of Siberian oil and gas. In the wake of the 1973 energy crisis, the United States, like western Europe, began to look to the USSR as an alternative source of natural fuels. The Soviets, in turn, discovered an added incentive to trade with the United States that led them to score one of the more striking commercial coups of the decade—the "great grain robbery."

Soviet leaders had long recognized the usefulness of importing Western technology, but it was only in the 1970s that their previously abundant supply of grain began to diminish dramatically, and their long-term projections predicted continued shortages. In 1972, when an unusually severe winter and late thaw caused the loss of 30 percent of the Soviet grain harvest, Moscow decided to approach the United States. By negotiating separately and secretly with the Department of Agriculture and individual grain producers, and by refusing to reveal to any of the individual sellers the total amount of grain they intended to purchase, the Russians managed to buy 19 million tons, a full quarter of the total American production, at a price ranging from $1.61 to $1.63 per bushel. That massive purchase pushed the price of grain for American consumers up to $2.43 per bushel, and one year later to $5 per bushel.[52] The Soviets were to display similar business acumen in negotiating the Euro-Siberian pipeline deal.

Between 1971 and 1974, U.S. exports to the Soviet Union rose substantially, from $144 million to $747 million, with a 1973 peak of $1.3 billion thanks to the previous year's grain deal. Meanwhile, western European trade with the East also continued to thrive, in part as a result of the new liberal atmosphere engendered by the U.S.-Soviet détente. During 1971–74, French exports went from $313 million to $718 million, while German exports shot up from $404 million to $1.8 billion.[53] In return, the Soviets began exporting substantial quantities of oil and natural gas to western Europe.

Even as Nixon and Kissinger were building the new trade relationship between Washington and Moscow, a wrench was thrown into the machinery that ultimately led to the collapse of the trade partnership. The 1972 agreement between the two governments had been subject to congressional approval; once it reached the Senate and the House as part of the 1973 trade reform bill, the accord encountered strong opposition from Sen. Henry Jackson, who sought to tie MFN status to a liberalization of Soviet emigration laws, particularly those regarding Jewish emigration. The senator was apparently motivated by the experience of being one of the first American soldiers to enter the

Bergen Belsen concentration camp as a liaison officer with the British at the end of World War II. The overwhelming impact of what he witnessed left a deep impression and made him particularly sensitive to the plight of the Jewish people throughout the world. More cynical observers saw in Jackson's stance on emigration an attempt to court Jewish support for his 1976 presidential campaign.

Whatever his reasons, Jackson proved a formidable adversary of the trade agreement. A three-way struggle involving the senator, the administration—which argued that the emigration question should be handled through quiet diplomacy—and Moscow ensued. The White House's power had diminished in the wake of the Watergate scandal, which partly explains Jackson's ability to overcome Kissinger's opposition to his initiative. The senator and the secretary did finally come to a compromise, which the Soviets seemed willing to accept. In return for MFN status, Moscow would informally pledge to allow 60,000 people emigrate a year but would make no public statement to that effect. But when Jackson trumpeted his victory to the media, the Soviets disavowed the deal.[54]

The trade agreement that eventually emerged from Congress in 1974 was weighed down by the Jackson amendment, which reads, in part:

> No non-market economy shall be eligible to receive Most Favored Nation treatment or . . . credits . . . [while it] . . . a) denies its citizens the right or opportunity to emigrate, or b) imposes more than a nominal tax on emigration.[55]

By denying the USSR MFN status, the Jackson/Vanik amendment—as it is commonly known—ensured in effect that tariffs on goods imported from the Soviet Union would be about ten times higher than those applicable to goods from other countries. Still, the amendment was a largely symbolic blow to the trade agreement. Even with MFN status, Soviet exports to the United States were unlikely to prove overwhelming. Another amendment, though, did more concrete damage to the agreement. Sponsored by Sen. Adlai Stevenson, Jr., it limited Export/Import Bank credits to the Soviet Union to $300 million for the three-year duration of the trade agreement. No more than $40 million of that amount could be used to purchase technology for the development of the Soviet energy sector, precisely the type of equipment the Soviets most wanted to acquire.[56]

Sen. Daniel Inouye explained congressional opposition to increased trade with the Soviets at this crucial juncture:

> The covert grain purchases of 1972, the domestic policies in the Soviet Union, difficulties in the arms limitation talks, and possibly questionable sales of high technology goods to the Soviets have all contributed to disillusionment and doubt about the desirability of increased U.S. trade with non-market countries.[57]

President Nixon signed the trade agreement into law, complete with the Jackson/Vanik and Stevenson amendments. In January 1975, the Soviets said they could not accept its provisions demanding guaranteed emigration and imposing credit ceilings, on the grounds that these amounted to interference in their domestic affairs. The 1972 agreement between Washington and Moscow, signed by Nixon and Brezhnev, was thus nullified. Once again, Congress had derailed a pro-trade initiative on the part of the executive, only this time, the disavowel of trade helped to derail the overall détente strategy. Kissinger's conception of détente was based in part on the trade pillar. While congressional restraints on commercial relations cannot be held directly responsible for the failure of détente, it is clear that by removing the trade pillar, Congress placed the entire weight of détente on politics. As the political relationship between Moscow and Washington soured once again in the late 1970s, the unstable support that remained was not strong enough to prop up détente.

After 1975, trade between the United States and the USSR actually expanded, but this was due almost exclusively to Soviet purchases of U.S. agricultural products. In October 1975, the two countries entered into a five-year grain agreement, with the parties undertaking to sell and buy 6 million tons of grain per year, respectively. By 1979, agricultural goods constituted nearly 70 percent of American exports to the USSR, while capital goods and other manufactured products accounted for just 15 percent of exports.[58]

Throughout the mid-1970s, trade between western Europe and the East grew steadily. The primary catalyst remained economic incentive; however, strategic considerations came into play as well. German *Ostpolitik* of the Brandt variety was continuing as before. In France, the Gaullist legacy remained potent, for both President Georges Pompidou and President Valery Giscard d'Estaing strove to preserve national independence by diversifying external relations. Giscard in particular expressed great interest in trade's potential for improving relations between France and the Soviet Union, and eventually peacefully transforming the USSR into a more benevolent society.[59]

Giscard's trip to Moscow in 1977 and Soviet leader Leonid Brezhnev's visit to Bonn in 1978 helped to hatch further commercial agreements. One outcome of the Brezhnev journey was that German firms contracted to build a number of large-scale industrial plants in the Soviet Union, and German banks undertook to finance the projects with mid-to-long-term loans at low interest rates.[60] These deals served as experimental blueprints for the Siberian pipeline project. Despite attempts in western Europe to establish trade agreements at an EEC-CMEA level, most arrangements remained bilateral and involved intensive competition among Western industries for access to Eastern markets. The energy sector in particular, and related industries such as steel, benefited from the increase in allied trade with the East, which the United States, by and large, tolerated without protest, but from which its own business community generally abstained. The west Europeans enjoyed the opportunity to develop the growing Soviet and east European markets as their restricted hunting preserve.

As tension between the United States and the Soviet Union increased again toward the end of the decade, the Soviets found an added political advantage in trade with the Europeans. Moscow hoped to weaken the solidarity of the alliance by increasing economic ties with key members of NATO—a mirror image of Western attempts to promote polycentrism in eastern Europe. The Soviets reasoned that the European economic incentive would clash with American strategic plans over the question of trade.[61] In retrospect, the Soviets demonstrated a great deal of farsightedness.

The rift within the West over trade policy widened markedly from 1978 on. The Carter administration, encountering a series of political crises with the Soviets, attempted to use economic sanctions in response. These necessitated allied cooperation to be effective. As the decade came to a close, the Europeans became less and less willing to sacrifice their economic interests for the sake of American foreign policy.

Carter first imposed sanctions in 1978, following the conviction of Aleksandr Ginzburg and Anatoly Shcharansky, two prominent Soviet dissidents, for "anti-Soviet activities."[62] The president, who had given his entire foreign policy a strong moralistic content, reacted to what he perceived as a serious expression of contempt for human rights by restricting the export of energy-related equipment to the Soviet Union.

The first major crisis with alliance-wide implications centered, however, on the Soviet invasion of Afghanistan in December 1979. The administration decided to show American displeasure through a number of actions. The first and most controversial move was the embargo on the export of American grain to the Soviet Union in excess of the 8 million tons still guaranteed under the 1975 agreement. Carter also suspended Soviet fishing rights within U.S. territorial waters, tightened restrictions on the sale of high technology, and called for a boycott of the 1980 Summer Olympic Games that were to be held in Moscow. Within a year, U.S. trade with the USSR was cut in half.[63]

To emphasize their disagreement with Carter's approach, the Europeans imposed no unilateral trade restrictions following the invasion of Afghanistan, agreeing only to grant no exceptions to the standing COCOM embargo list.[64] The EEC, along with Canada and Australia, did pledge not to undercut the grain embargo by holding their sales to the Soviets to "normal" levels. The allied definition of "normal," however, proved sufficient to nullify whatever effect the embargo might have had, since it permitted the Soviets' suppliers to compensate for most of the embargoed U.S. sales.[65] Nor did the Europeans go along with the American embargo initiative in other areas. While overall U.S. exports to the Soviet Union dropped 54 percent from 1979 to 1980, those of its major allies increased by 35 percent (France), 32 percent (West Germany), and 67 percent (Great Britain).[66]

In the wake of Afghanistan, sentiment within both the executive and the legislative branches of the American government had strongly veered toward anti-trade.[67] The Reagan administration began where the Carter team had left off: Reagan made it clear from the outset that trade with the East would be dictated primarily by security considerations, and not by economic benefit to American exporters. There was, however, one outstanding exception: agriculture. Reagan lifted Carter's embargo on grain to make good on a campaign pledge to American farmers. Then, he instructed Secretary of Agriculture John Block to negotiate new, long-term grain accords with the Soviets. This development prompted the Europeans to complain bitterly that Washington was willing to sacrifice European economic interests, but not its own, in the name of security.

Ignoring this inconsistency, the White House tried to convince the allies of the need for trade restrictions in three broad, but related, areas: technology transfer, credits, and energy. One tactic the United States employed to get the allies to restrain trade was to advocate cuts in Western credits to the East. Fewer credits on more stringent terms would make energy deals and sales of Western technology difficult. In January 1980, the administration asked OECD members to cut in half all official and government-guaranteed credits designated to finance trade with the USSR. The allies declined to do so.[68]

At the same time, the "atmospherics" of the superpower relationship were worsening. Harsh anti-Soviet rhetoric emerged with increasing frequency from Washington, while the level of anti-American discourse in Moscow rose. The alliance itself was experiencing strain as the NATO decision to deploy Pershing II and cruise missiles in response to the Soviet installation of an arsenal of SS-20s pointed westward aroused a strong protest in Western Europe.[69]

By 1981, trade between East and West dropped by nearly $10 billion.[70] While the cause for the decrease in U.S.-East bloc trade had been primarily political, western European-East bloc commerce had slackened for economic reasons. The general economic recession led to a 2 percent decline in overall world trade by 1982, a phenomenon that did not leave East-West trade unscathed. The mounting eastern European debt made Western financial institutions wary of furnishing new credits and rolling over existing ones as they had done in the past, especially without government-provided default insurance.

Between 1974 and 1980, the net liabilities of the East bloc countries jumped from $6.7 billion to $45 billion. A year later, while the USSR owed the West $10.2 billion, the rest of the CMEA owed a staggering $48 billion.[71] The East bloc, especially Poland and Romania, was beginning to resemble certain Third World countries in the magnitude of its indebtedness to the West. The resulting dilemma was also much the same—a vicious cycle. By refusing additional loans, European and American banks made it more difficult for the outstanding debt to be repaid or even serviced, since all new credits were needed to finance the development of additional Eastern exports upon which ultimate repayment and service were to be based.[72]

Despite these constraints, western Europe's trade with the East remained much more significant than America's throughout the early 1980s. The figures, though, can be misleading. The EEC accounted for more than 60 percent of world trade with the East bloc, compared with just 6 percent for the United States. As a percentage of its own foreign trade, however, European commerce with the East, while far greater than America's (1.14 percent), was not overwhelming: the overall EEC figure was 4.14 percent, with West Germany (6.20 percent) and France (4.20 percent) at the top of the list. Nor was trade with the East as a share of European GNP strikingly high. EEC-CMEA trade as a percentage of the Community's GNP was 2 percent, while trade with the CMEA was 1.58 percent of France's GNP and 2.90 percent of West Germany's GNP, compared with only 0.2 percent of America's GNP.[73]

Most of western Europe's trade with the East involved a small number of very important companies and industries, some of which produced advanced technology, a sector vital to the future growth of European economies. Even if the contracts with the Soviet bloc represented a small percentage of each company's total business, they could still prove crucial on the margin, making the difference between a poor year and a good one, and allowing the financing of more product research and development. Other industries deeply interested in East-West trade were the more classical areas of production such as steel, which was suffering greatly from the recession. Here again, even marginal trade with the East could have a tremendous impact on financial performance.[74] According to calculations by the Office of Technology Assessment in Washington, 92,000 West German jobs—0.4 percent of the German work force—depended on trade with the Soviet Union and 220,000 jobs on commerce with the East bloc as a whole.[75]

The western Europeans thus found themselves in a paradoxical position. They had to fend off criticism from Washington that their economies were dangerously dependent on trade with the East. Yet they also had to show that commerce with Communist countries was important to them for economic reasons, and that curtailing it would place an unfair and unnecessary burden on their economies. The European position does seem justified. Cutting off trade with the East would hardly cripple the EEC; however, it could do some serious damage to specific and important economic sectors, especially the increasingly vulnerable smokestack industry. The cost could be borne if required, but it involved serious sacrifices in difficult economic times. As unemployment grew in a context of worldwide recession, such "sacrifice" could fuel social and political unrest, and even anti-American sentiment within formerly stable members of the alliance.

By 1982, the United States and its allies were as far apart on the issue of trading with the East as they had been at any time since World

War II. In every alliance organization, from NATO to COCOM to the OECD, Washington attempted to push the Europeans toward restraining commercial relations with the Soviet Union and its satellites. The Europeans refused to give ground. At the very time when the United States was returning to a philosophy of East-West economic confrontation, European and Soviet negotiators were shuttling between Moscow and Western capitals, constructing what seemed destined to be the largest East-West commercial undertaking ever: the Siberian pipeline.

6

The Pipeline Embargo

The United States had been uneasy about the Siberian pipeline project from its inception. Carter administration officials warned their counterparts in Europe of American qualms concerning potential Soviet energy leverage over the EEC. At the same time, administration proponents of selective trade denial to the East argued that the Soviet energy sector would be particularly vulnerable to economic pressure. Following the arrest of several prominent Soviet dissidents in 1978, Carter placed restrictions on the sale of oil and gas technology to the USSR. When the Soviets invaded Afghanistan the next year, some officials contemplated an embargo on technology that would have included equipment for the pipeline.

The Europeans in turn made it clear to Washington that they would not abandon the pipeline project. They also tried to assuage American fears about their eventual dependence on Soviet gas. The allies' commitment to the pipeline apparently convinced Carter not to take further action. In any event, the lingering hostage crisis in Iran diverted Washington's attention from the problem.

Ronald Reagan's election to the presidency made the pipeline an issue for alliance debate once again. The new president and the leading members of his cabinet, notably Defense Secretary Caspar Weinberger and Assistant Secretary Richard Perle, adopted a markedly hard line approach to relations with the Soviet Union. They believed that American weakness over the previous half decade had allowed the Soviets to achieve military parity with the United States and that increased trade had only served to help Moscow overcome its technological backwardness.

The Reagan administration embarked on a military buildup and called into question the West's commercial ties with the East. As officials studied the trade problem, a three-pronged policy emerged: to slow down both legal and illegal technology flow to the East; to avoid providing massive credits and preferential terms of trade; and to use trade as a tool to wring political concessions from Moscow, and possibly even hasten positive change within the Soviet Union. Such was Washington's line of argument when Reagan met with the heads of state of America's major industrial allies at Ottawa in July 1981.

During the Ottawa summit, the American delegation made a point of requesting that COCOM adopt broader export restrictions and that the European allies limit low interest rate credits to the Soviet Union. The West, it was argued, should not enter into financial transactions with the East on more favorable terms than Western countries give one another.[1] The Americans maintained that the arrangements for the pipeline were a case in point. U.S. officials also raised the issue of European energy dependence on the USSR, and expressed concern that the pipeline would bring the Soviets a hard currency windfall and significantly increase their energy production for military and domestic purposes.[2] Despite these criticisms of the pipeline, Washington seemed most interested in the larger issue of restraining future credits to the East.

If anything, the summit was one of a series of events that reinforced the Europeans' impression that the United States would not, in the end, oppose completion of the gas project. Several months before the Ottawa summit, the NATO Council of Ministers had agreed that one ally would not employ sanctions that might undermine the position of another in East-West trade negotiations or contracts.[3] This would seemingly deter the imposition of unilateral export prohibitions.

Then, in the fall of 1981, the State Department sent a group of officials, led by Under Secretary of State for Economic Affairs Meyer Rashish, to Europe to discuss alternatives to the pipeline. The delegation proposed a combination of American coal, synfuels, and nuclear power technology in place of Soviet gas. However, as soon as the talks turned to specifics, it became clear that the substitute plan was impractical. In order to make the sale of coal to western Europe economically feasible, a substantial port modernization program in the United States would have been necessary. This was impossible at a time when the administration was going to great lengths to cut the federal budget. As for synfuels, they were not even being developed for the U.S. market. In the near term, nuclear power was also out of the question. The Carter administration, in its effort to fight the proliferation of nuclear weapons, had made it difficult to export such technology. Moreover, its development, while advanced in several EEC countries, notably France, posed serious political and environmental problems. The Rashish team is also reported to have suggested to the incredulous Europeans that they import more OPEC oil in the place of Soviet gas.[4]

The bottom line was that none of the alternatives to the pipeline advanced by Washington were as cheap and feasible as Soviet gas. Nor could they match the Euro-Siberian project in promoting jobs and business in western Europe.[5] As negotiations over the various pipeline contracts reached a conclusion in late 1981, even some American officials opposed to the deal, such as Secretary of State Alexander Haig, considered the project a fait accompli.

ROUND ONE: THE SEARCH FOR COMPROMISE

On December 13, 1981, the crisis that had been brewing in Poland for months finally erupted. The Soviet-backed military government of Gen. Wojciech Jaruzelski declared martial law and imprisoned leaders of the Solidarity union. President Reagan, stating that the USSR bore "a heavy and direct responsibility for the repression," decided on December 29 to impose economic sanctions on the Soviets. These included a suspension of Aeroflot's landing rights in the United States, a freeze on negotiations for a new long-term grain agreement, the closing of the Soviet Purchasing Commission, and a halt to the issuance of export licenses for computers and electronic equipment.[6]

The president also announced that henceforth, special export licenses would be required for an expanded list of oil and gas equipment, and that those licenses already awarded would be immediately suspended until martial law was lifted, Polish political prisoners released, and talks among the government, Solidarity, and the Roman Catholic Church commenced.[7]

From the outset, confusion shrouded what proved to be but the first round of pipeline-related sanctions. Some administration officials believed that Reagan's edict was not retroactive; thus, pipeline equipment contracts already signed were not to be affected by the sanctions: "I never imagined that the President's decision would be retroactive," former Secretary of State Haig has written. "I doubt this was the

President's intent, either; certainly, it never won support around the National Security Council table."[8] But the Commerce Department, responsible for export control administration and eager to show Defense Department critics that it was not "soft" on trade with communist countries, interpreted the sanctions as indeed being retroactive. Consequently, several American firms, including General Electric, Caterpillar, Dresser, and Cooper Industries, had to cancel contracts to sell the Soviets equipment destined for use on the Euro-Siberian pipeline. They were also prohibited from supplying their products to European licensees who were contractors for the project, even if they had already agreed to do so.

Several European companies found themselves in a serious bind. AEG-Kanis of West Germany, John Brown of Scotland, and Nuovo Pignone of Italy all relied on General Electric rotors, nozzles, and stator blades for the 125 turbines they were building for the pipeline. Each of the pipeline's 41 compressor stations required three turbines: one for daily use, one for reserve, and a third for periodic maintenance operation. Because GE had shipped 23 rotor sets to the European companies before the ban, it was technically feasible for 23 of the 41 compressor stations to function. This meant the pipeline could operate at a 70 percent throughput rate, but without backup or maintenance turbines.[9] Gas thus could flow to the Soviets' Western customers in the immediate future, but long-term concerns about maintaining the pipeline in continuous operation and the fulfillment of binding contractual obligations necessitated the supply of the remaining turbines.

Since the GE parts were no longer available, the Europeans had to find another, non-American supplier. Two options existed, each with significant drawbacks. Alsthom-Atlantique, a state-controlled French firm, was the only one of GE's manufacturing associates to possess a license to make rotor kits itself. Alsthom had in fact contracted to supply the Soviets with 40 spare rotor sets in 1984. Yet Alsthom had never actually produced a rotor; it still lacked the tools and the know-how. In order to take up GE's slack, it would have to build a new factory or retool an existing one. Noting "the obvious political risks" involved in accepting additional orders for the pipeline and fearing it would jeopardize its relationship with GE, Alsthom hesitated to take the place of its American licensor.[10]

Rolls Royce of Great Britain also made rotors with the 25 Mw capacity that the pipeline planners desired. But using the Rolls Royce model—allegedly less reliable than GE's—would have necessitated the redesign of the turbine unit. That would, in turn, have precluded several of the turbine producers from participating in the project. It is also doubtful that Rolls Royce could have supplied the rotors in time to keep the pipeline on schedule.[11]

With these constraints in mind, the European companies adopted a wait-and-see approach with the thought that by the time all the compressor stations were needed for the pipeline, the American sanctions would no longer be in effect. The diplomats, they hoped, would sort things out.

A diplomatic settlement of the stalemate became less and less likely during the first months of 1982. The Americans were angered by the Europeans' refusal to participate in economic sanctions against the USSR. Both West Germany and France had reasserted their will to continue ongoing commercial projects with the Soviets, and West German Chancellor Helmut Schmidt had not even bothered to interrupt an official visit to East Germany when martial law was declared in Poland.[12] A month after the Polish crackdown, Gaz de France signed its supply contract with the Soviets, followed a few days later by the German pipe manufacturer Mannesmann. On January 28, Italy announced that it had agreed "in principle" to buy gas from the pipeline.

A succession of American officials publicly voiced their criticisms of Europe's economic involvement with the East. Assistant Secretary of Commerce Lawrence Brady denounced the influence of a "Soviet lobby" in the heart of Western business, and warned that Moscow's "energy diplomacy" sought to place western Europe in a situation of dependence on Soviet natural gas.[13] Stephen Brian, deputy under secretary of defense for trade and security policy, declared that "certain European banks could find themselves in a difficult position" if they continued to grant credits to the Soviets.[14] Meanwhile, Secretary of Defense Weinberger advocated an extension of the American trade embargo to a virtual financial blockade of the Soviet Union and its allies by urging that the West withhold further credits and refuse to reschedule East bloc debts.[15]

The United States also called the first ministerial-level meeting since 1958 of COCOM, the inter-alliance organization that oversees export controls, to discuss that institution's effectiveness. During the talks, the Americans urged a broadening of the norms used to define strategic exports in COCOM and attempted to engage the Europeans in a debate over the pipeline.[16] Both within and outside COCOM, U.S. officials told their European counterparts that Washington was considering widening the pipeline sanctions to include products manufactured abroad under American license, and to ban any participation in the project by "off-shore" subsidiaries of American companies, even those using no U.S.-origin technology.[17] To many Europeans, the United States seemed to be declaring outright economic war not only on the Soviet Union and its east European allies, but even on the NATO countries if they stood in the way.

As the United States and the Europeans continued to bicker over East-West trade issues in general, and the pipeline in particular, Haig sent Under Secretary of State for Security Assistance James Buckley to Europe in an effort to elicit more cooperation from the allies on economic policy toward the East. According to Haig, Buckley was told to press the Europeans to restrict credit, and to promise them American cooperation on the pipeline in return.[18] Such a trade-off seemed to have possibilities for resolving the growing rift within the alliance. Other sources maintain that Buckley was also told to threaten the Europeans with an extension of the sanctions to the foreign subsidiaries of American firms manufacturing pipeline equipment unless credits were limited.[19] Buckley insists that he and his team "only discussed the issue of credits."[20]

Just as the Buckley mission began, Washington won a victory on the credit issue within the OECD. In theory, interest rates charged to countries outside the OECD by official lending institutions follow guidelines known as the "Consensus," established in 1976. The rates vary depending upon whether the country receiving the loan fits into one of three categories: rich, middle income, or poor. In 1976, most East bloc countries were placed in the middle category, and were granted interest rates between 7.8 and 11 percent, as opposed to 11 to 12.4 percent for rich countries. In February 1982, the Consensus agreed in principle to reclassify a number of countries, including the Soviet Union, which was placed in the rich category.[21] This change—which was not officially approved by the participating countries until after the Versailles summit—was due in part to mounting American pressure on the EEC to grant the East bloc less favorable terms of credit.

However, the Consensus applies only to loans made through official government export credit programs, not to those made by private banks. Because a great deal of European financing for East-West commercial projects is private, the reclassification did not eliminate low interest loans to the USSR.[22] German financing for the pipeline, for example, was entirely private. In addition, even the higher rate of 12.4 percent remained below the interest rate then prevailing in France.

The Americans were as concerned with limiting the volume of credits as they were with seeing that a higher rate of interest was applied. Despite the OECD agreement, the Buckley team took an aggressive anti-pipeline stand, according to some of the Europeans with whom it met. Wilhelm Christians, president of the Deutsche Bank, was one of the few private citizens to be lobbied by the Buckley mission. He claims that the American envoys were on "an anti-trade crusade" and displayed "little understanding of the historical situation of western Europe." Christians, an ardent proponent of East-West trade who negotiated the first German-Soviet gas deals in the late 1960s, adds that the Americans "seemed to think we were ignorant of the Soviet threat, which is strange considering Washington is 5,000 miles from the Soviet tanks, and from here in Germany, the distance is only 200 miles."[23] Buckley says that his reception in Europe was characterized by "a variety of enthusiasms."[24]

Washington wanted the Europeans to renegotiate the pipeline contracts to bring the terms of credit in line with the new OECD Consensus. The Europeans made it clear they did not believe that the OECD agreement was retroactive; the pipeline contracts had been signed before reclassification, and thus the old Consensus rates should apply. Besides, the German financing clearly was not affected by the change in OECD rules because it was entirely private. If Buckley could not get the Europeans to rework the contracts, State Department officials say that he at least hoped to get a pledge from them that nonofficial loans in the future would offer terms of credit less favorable to Moscow, and that loans of both kinds would be less abundant. And if this proved impossible, the State Department apparently told Buckley to demand European participation in a broad reassessment of East-West trade policies.

In May 1982, the Buckley mission was still in Europe. Some American officials insist that had the delegation been given more time, it would have succeeded in averting the crisis over the pipeline. Robert Hormats, then assistant secretary of state for economic and business affairs, argues that "Buckley was very close to achieving an understanding with the Europeans that involved credits and the pipeline. But then it was June and the Versailles Summit was upon us. Everything became highly politicized and too many issues were intertwined."[25]

Haig, in preparation for the summit, had dispatched George Shultz, then president of the Bechtel Corporation, on a round-robin mission to Canada, France, Italy, Japan, Great Britain, and West Germany to discuss the East-West trade problem. Shultz reported that the Europeans might be willing to restrain cheap credits to the East.[26] As Versailles approached, Haig decided to present the Buckley trade-off offer to his European counterparts, but in the form of a veiled threat:

> I made it plain to the European foreign ministers that if we did not at least have progress on a cooperative policy to limit future . . . credits to the Soviet Union, the United States would find it difficult not to apply sanctions that would prevent the use of American technology [including licenses and patents] for the pipeline.[27]

The Europeans continued to hold out.

On the eve of the summit, Haig thought he had found a workable compromise. The French were desperate for the United States to intervene in foreign-exchange markets to bolster the weak franc, which threatened France's economic recovery. The economic policy of the Socialist-Communist coalition in power was failing, and the French were facing an increasingly difficult situation under the "Snake," the EEC's monetary system. With the Deutsch mark growing ever stronger and the franc ever weaker, the French found themselves backing into another humiliating devaluation of their currency. The Quai d'Orsay hinted that an American decision to come to France's rescue would prompt France and the other European allies to limit credits and favorable terms to the East, provided Washington also agreed not to hinder European participation in the pipeline. Haig urged the president to offer the trade-off to the Europeans at the summit.[28]

But the White House was reluctant to go along with Haig's initiative. Treasury Secretary Donald Regan opposed the plan. Government interference with currency floatations was contrary to his and the administration's noninterventionist philosophy. Weinberger and National Security Adviser William Clark were strongly opposed to the pipeline as a matter of principle and did not want to "trade" it for an agreement on credits. They also believed that Soviet economic difficulties could be greatly exacerbated by the broadest possible trade embargo on Western technology. Stopping the pipeline would serve this policy as well.

Clark told Haig that a decision on extending the pipeline sanctions to subsidiaries of American companies and foreign firms using U.S. technology would be put off until a National Security Council meeting after the summit.[29] This did not prevent the secretary of state from proposing his compromise to the European foreign ministers and finance ministers the night before the summit plenary session:

> Very late, we reached an understanding: the United States would bolster the franc on a case-by-case basis in return for restraint by the other governments on future credits to the Soviet Union. This ministerial agreement, if sustained by the heads of state at the plenary, would by implication trigger a decision favorable to the Europeans on the pipeline issue.[30]

During a vigorous debate at the plenary, it appeared that several of the leaders involved, including French President François Mitterrand, were not aware of the compromise worked out by their ministers. This was not entirely surprising. The Falkland Islands war and the Israeli invasion of Lebanon dominated the summit and diverted attention from the East-West trade issue. The proposal was, however, finally understood, and apparently accepted, by all. The closing statement of the summit contained a declaration by the allies that the West would restrain subsidies to the East. The wording of the accord was vague, the participants simply agreeing to adopt "commercial prudence in limiting export credits."[31] No quantitative ceilings on credit or subsidies were included, but Washington had, in principle, achieved its goal. While the pipeline was not mentioned in the final communiqué, it was understood, according to Haig, "that the United States would not apply retroactive, extra-territorial pipeline sanctions."[32] Thanks to last minute maneuvering, and Haig's willingness to ignore White House directives, it seemed the pipeline case had finally been resolved.

ROUND TWO: THE FAILURE OF DIPLOMACY

The alliance entente on East-West trade lasted just 24 hours. Treasury Secretary Regan, apparently unaware of the summit agreement, told the press that Washington would not intervene in foreign exchange markets to support the franc. The French were outraged and promptly disclaimed their part of the summit "deal." President Mitterrand declared that his country "would not support economic warfare against the Soviet Union and is not bound by the summit declaration to cut the amount of credit extended to Moscow." German Finance Minister

Manfred Lahnstein, angered as well, stated, "We will continue to work with Eastern European countries and the Soviet Union as usual."[33]

In Washington, National Security Adviser Clark and Defense Secretary Weinberger, who had opposed the Haig trade-off to begin with, were anxious to react to the Europeans' repudiation of the summit accord. At a National Security Council meeting that some State Department officials say was scheduled so that neither Haig nor U.S. Trade Representative William Brock could attend, Clark and Weinberger urged the president to extend the pipeline sanctions. Haig, whose deputy Larry Eagleburger was present, claims that Reagan was shown "only the strongest option paper."[34] Former Assistant Secretary Hormats concurs: "They put one over on him [the President]. He wasn't fully informed of what the options were."[35]

On June 18, Reagan announced that he was extending the sanctions to American subsidiaries based abroad and foreign companies producing equipment for the pipeline under license from American companies. This round of sanctions was clearly intended to be retroactive. Thus, even the 23 GE rotor sets already shipped to Europe were subject to embargo. It also meant that any American-designed production processes or technologies, even those used by western European companies on western European soil to produce anything connected with the pipeline, were banned.[36] Finally, even if the technology they were using to produce equipment for the pipeline was not of U.S. origin, the European subsidiaries of American companies were prohibited from taking part in the project.

A statement released by the White House attributed the extension of the sanctions to the president's desire to "advance reconciliation in Poland" by continuing economic pressure on that country and on the USSR.[37] Despite the reference to Poland, it is clear that Reagan was equally concerned with preventing completion of the pipeline. "The Polish crisis provided a convenient pretext for dealing with the pipeline issue, which had long nettled their [Defense and NSC's] strategic sensitivities," argues Haig.[38] At the same time, the president became convinced that the Europeans could be persuaded to follow American directives on the larger trade issues such as credit only by a dramatic gesture. Richard Pipes, former national security adviser for soviet affairs, puts it frankly that, "We saw no reason in continually deferring, as had been done in the past, to the Europeans. We always humor them, they never humor us. Enough was enough."[39] The pipeline thus became both a target of American policy and a means to assure European compliance with Washington's restrictive East-West trade strategy.

The new round of sanctions was in large measure responsible for Haig's resignation from the cabinet on June 25. He was replaced by George Shultz. More important, the American initiative met with vehement protest from the Europeans. The EEC issued a statement decrying "the state of extreme political and economic tension with Washington." Bonn pronounced itself "dismayed" at the sanctions and called them "a contradiction to what was agreed and discussed at the economic summit." EEC trade negotiator Sir Roy Denman said, "Our trade relations with the United States are the worst I have seen since the end of the war." The normally conservative *Financial Times* of London wrote that "the components of the Western Alliance are coming apart."[40]

Officials in Washington admit that they failed to anticipate the violent European reaction. Perhaps the allies' restrained demeanor following the first round of sanctions lulled Washington into believing it could succeed in extending its embargo. Moreover, the administration believed that France's Mitterrand and British Prime Minister Margaret Thatcher ultimately shared Washington's views on East-West relations and would eventually adopt a tougher trade policy. The French president, a socialist, had proved to be a committed "Atlanticist" when it came to dealing with the Soviet Union—a stauncher and more reliable ally than his conservative predecessors Giscard, Pompidou, and De Gaulle. Thatcher was also an ardently faithful ally of the United States. But Washington's sanctions pushed both leaders too far.

The White House had received fair warning from the State Department and foreign service officers stationed in western Europe that extending the sanctions would seriously alienate the European allies. "We told Washington that this was a bad place to pick a fight with the Europeans, no matter how justified our arguments," said Michael Ely, a senior American foreign service officer then serving in Paris.[41] Haig, who opposed the pipeline for strategic reasons, understood that the Europeans had a lot at stake, and that "it was, quite simply, too late to say no."[42]

In the end, it was not just the fact that the pipeline would provide an alternative source of energy to the EEC, or that it would ensure tens of thousands of badly needed jobs in Europe, that provoked the allies' fury. More fundamentally, the European response was one of gut nationalism from countries trying to assert their independence from American foreign policy directives. The second round of sanctions touched a raw nerve because Washington was giving orders to European owned, operated, and even state-controlled companies. As Rudolf Augstein, a leading West German political commentator and publisher of the influential weekly *Der Spiegel*, notes, "The Americans were treating us as if we were not sovereign states. We could not sit still and let them run our lives for us."[43]

The strongest and most succinct expression of European opinion came from one of Washington's best friends. A former member of the London Bar, steeped in the tradition of the Common Law, Margaret Thatcher's denunciation of the sanctions was couched in fundamental legal terms. What offended her most was their extraterritorial and retroactive nature. "It is wrong," she told the House of Commons, "for one very powerful nation" to try to prevent the fulfillment of "existing contracts that do not, in any event, fall under its jurisdiction."[44] Even though Thatcher was at least sympathetic to some of Washington's strategic concerns regarding the pipeline and the credit issue, Great Britain—like the rest of the alliance members—refused to be considered an American vassal state.

The European pipeline suppliers, almost all of whom utilized American technology in manufacturing their products or were affiliated with American companies, were faced with a bitter dilemma. They could attempt to shift to non-American technology, but that would have entailed the drawback mentioned previously of having to redesign the entire apparatus in order to accommodate the different technology. They could try to persuade Moscow to develop some of the technology in question itself. Although the Soviets claimed they were capable of manufacturing the necessary 25 Mw turbines, Western experts were skeptical. The USSR did have the ability to build 10 Mw turbines. Two of these could be substituted for a single 25 Mw machine. However, this would mean more frequent maintenance, higher operating costs, and lower efficiency.[45] Finally, the Europeans could defy the American export controls and ship the equipment already in their possession to the Soviet Union—a move that the pipeline contracts effectively required under a clause that allowed the Soviets to demand a penalty of up to 5 percent of the contracted equipment's value for nonfulfillment of supply obligations.[46] The USSR reminded the Europeans of this clause by calling representatives of the supplier companies to Moscow in July and informing them that Moscow would not hesitate to insist upon penalty payments in the event of delivery delays.

Once again, the companies decided to wait for a diplomatic resolution to the problem, or orders from their respective governments. And once again, a quiet compromise seemed increasingly unlikely. Washington continued to pressure the European manufacturers with a range of threats, including penalties, criminal prosecution, and "blacklisting" of all of their exports to the United States market if they defied the embargo. Most of the White House's attention was directed at Alsthom-Atlantique. The administration knew that without Alsthom's GE-patented rotor sets, the other pipeline equipment would be of little use to the Soviets, at least in the short term. By convincing Alsthom to respect the embargo, Washington intended to complicate the project, thereby increasing each participant's costs to a point where continued involvement would, it was hoped, be prohibitive.

In an interview with the French newspaper *Le Monde*, American ambassador Evan Galbraith noted that the license contract between GE and Alsthom contained a clause allowing the United States to penalize Alsthom severely if it delivered oil and gas equipment to the East bloc without GE's accord. If Alsthom ignored the American embargo, life for that company in the United States could well "become a nightmare," he warned.[47] Adding to the problem, Alsthom's parent firm, the Compagnie Générale d'Electricité, was one of the largest French investors in the United States and feared retaliation if its subsidiary failed to heed Washington's wishes.

At first, the European governments moved cautiously. In late June, Thatcher asked the Ministry of Trade to consider applying for the first time the 1980 Protection of Trading Interests Act. This would allow British firms, including American subsidiaries incorporated in Britain, to disregard the sanctions as a threat to the U.K.'s commercial interest. The European Commission issued a statement in Brussels warning Washington that a "trade war" could result if the sanctions were maintained. But the EEC leaders took no other affirmative action throughout most of July, in the hope that negotiations with Washington would prove successful.

On July 30, Reagan outraged the Europeans even further, and ultimately triggered their decision to disregard the sanctions. The president announced that the United States and the Soviet Union had signed a new one-year grain deal. "The Soviet market," he said, "is the biggest in the world, and we want to recapture it after an embargo that allowed other countries, like Argentina and Canada, to take our place. Our national economy needs it."[48] The president was responding to a serious internal political problem. A grain glut and falling prices had forced American farmers to lobby Washington for greater access to the Soviet market.

As far as the allies were concerned, this was the height of hypocrisy. The grain accord made it seem that Washington believed that only the Europeans should bear the costs of imposing economic sanctions on the East bloc. Reagan's initiative also negated whatever impact the "hard currency" critique of the pipeline might have had. After all, argues Helmut Schmidt, "It was more than a little ridiculous for Washington to sell the Soviets grain and ask them to pay in hard currency while telling us that the pipeline was a strategic disaster because it gave Moscow too much hard currency."[49] The American act, and the failure to convince Washington to rescind the pipeline embargo, spurred the western Europeans to take the initiative.

Within a week, Lord Cockfield, Britain's trade secretary, invoked the Protection of Trading Interests Act, and the Germans soon thereafter adopted a similar measure. The French and the Italians, who had indicated just prior to Reagan's grain announcement that they expected their companies to honor contracts with the Soviet Union, took definitive action. Minister of Industry Jean-Pierre Chevenement signed an administrative decree ordering Dresser-France to deliver its compressor equipment. The Italian government told Nuovo Pignone to fulfill its turbine supply contract.

In open violation of the American embargo, the first shipments of pipeline equipment left European ports, bound for the Soviet Union, in late August. One ship departed the French port of Le Havre carrying three 57-ton compressors built by Creusot-Loire and Dresser-France, a subsidiary of Dresser-USA. Another vessel embarked from Glasgow, transporting six John Brown turbines built with GE parts. A third ship weighed anchor at Bremen, hauling two AEG-Kanis turbines, also made with GE parts.[50] In response, Under Secretary of Commerce Lionel Olmer immediately announced that the companies would be denied access to all American goods, services, and technology, and could be fined up to $100,000.[51] By October, a total of 12 European companies, including Mannesmann and Nuovo Pignone, were blacklisted by Washington for defying the embargo.

PATCHING THINGS UP

As summer turned to fall, the war of words—and to a lesser extent of deeds—continued across the Atlantic. Neither side seemed willing to yield. But in the United States, a growing number of policy makers were beginning to realize that the sanctions were a failure and that American diplomacy was facing a major debacle.

The CIA was convinced that the embargo should be lifted. In August, the Agency directed a secret report to the White House that stated:

> We believe that the USSR will succeed in meeting its gas delivery commitments to Western Europe thru the 1980s. Moscow has a wide range of options to accomplish this end. . . . It could commence delivery on schedule by using existing pipelines which now have an excess capacity of at least 6 bcm's/year. It could use some combination of Soviet and Western European technology to build the proposed line which should delay commencement of deliveries till no later than the end of 1985. At substantial cost to the domestic economy, the USSR could divert construction crews and compressor equipment from the new domestic pipelines under construction, or even dedicate one of those lines to export.[52]

Meanwhile, leading business lobbies, including the United States Chamber of Commerce and the National Association of Manufacturers, began to pressure Congress to take action. The "unprecedented blanket prohibition over US subsidiaries and affiliates and control of previously licensed US technology pose serious questions concerning the present direction of US international economic policy," the Chamber of Commerce argued.

In mid-August, Congress reacted. The House Foreign Affairs committee voted 22 to 12 to rescind the embargo. It underlined "the perception and the fact that the US controls are more of a sanction upon Europe than upon the Soviet Union." Hearings in the Republican-controlled Senate made it clear that most legislators in that body also believed the sanctions were futile. Martial law still reigned in Poland, and the embargo had not prevented the Soviets from obtaining the equipment they needed from western Europe. Moreover, the administration's policy was "driving a wedge between the United States and its allies, to what must be the obvious delight of the Kremlin," said Sen. Paul Tsongas. Rep. Henry Reuss, chairman of the Joint Economic Committee, which includes members of both houses of Congress, summed up the legislative branch's nearly united opposition to the pipeline embargo thus: These sanctions have "seriously undermined the trust and will to cooperate on which the alliance relies."[53]

Indeed, the central irony of the pipeline crisis was already clear. Allied nations had become adversaries, while the USSR, their common adversary, not only remained virtually unaffected but also, in a very real sense, was supported by the European members of the Atlantic Alliance in defiance of its most powerful member, the United States.

Secretary of State Shultz agreed with this assessment and gave top priority to resolving the crisis. Some kind of face-saving arrangement was needed, however, if the United States was to lift the embargo. At a meeting with allied foreign ministers at Montreal in October, Shultz asked his counterparts to undertake a full review of East-West trade relations. This was to include a reevaluation of COCOM standards and further discussions on credit and energy trade with the Soviets. In return, the secretary said, the Americans would drop the pipeline sanctions.[54] Ironically, one of the options the Buckley mission is said to have presented to the Europeans months earlier, before the Versailles summit, was just such a deal.[55]

Formal negotiations involving the United States, West Germany, France, Italy, Japan, Canada, and the European Community Commission began that month, with the goal of elaborating a framework of principles for future East-West trade. As the discussions progressed into early November, President Reagan adopted Shultz's idea of rescinding the pipeline sanctions and simultaneously announcing that the allies had agreed to discuss an overall economic strategy toward the USSR. Linking the two developments would make Washington's reversal less embarrassing. "It seemed reasonable enough to us that the Americans wanted something in return for dropping the sanctions," says an official in the British Department of Trade then in charge of the Soviet desk. "We were ready to play ball and give them the studies."[56]

When Soviet leader Leonid Brezhnev died on November 11, the White House decided that dismantling the embargo could also be disguised as a gesture of goodwill toward the new Soviet leadership. On November 13, in a radio address, Reagan lifted both sets of pipeline sanctions and announced an allied agreement on East-West trade policy. According to the president, the allies would take three immediate steps:

First, each partner has affirmed that no new contracts for the purchase of Soviet natural gas will be signed or approved during the course of our study of alternative Western sources of energy. Second, we and our partners will strengthen existing controls on the transfer of strategic items to the Soviet Union. Third, we will establish without delay procedures for monitoring financial relations with the Soviet Union and we will work to harmonize our export credit policies.[57]

The president barely mentioned Poland—which was ostensibly the cause for the sanctions in the first place. And he somewhat overstated the alliance accord, since the Europeans had simply agreed to participate in studies of energy trade, technology transfer, credit arrangements, and the security implications of East-West commerce while declaring a temporary moratorium on other gas deals with the USSR that did not affect the pipeline. Nonetheless, it seemed Washington had found a dignified exit from the embarrassing alliance crisis it had precipitated.

The French, however, unlike the British and the other members of the alliance, were not ready to play ball. They had maintained all along that, in view of the unilateral nature of the pipeline sanctions, the Europeans had no obligation to make concessions to Washington in order to have them lifted. Within hours of Reagan's address, Paris declared that it was "not party to the agreement announced this afternoon in Washington."[58] It took several weeks of careful diplomacy to undo this latest bit of damage to alliance solidarity.

Meanwhile, construction of the pipeline continued. On July 28, 1983, the Soviets announced triumphantly that the last length of pipe had been laid.[59] However, none of the compressor stations were on line. On New Year's Day 1984, Moscow said that natural gas had begun to flow through the pipeline. Western analysts, though, believed that only one of the 41 compressor stations was operational, and that the gas was moving through parts of an older pipeline during some stages of its journey westward.[60] Nevertheless, the Soviets insisted on claiming a propaganda victory against "the notorious policy of sanctions which became an inseparable element of the U.S. aggressive imperialist policy." Moscow further boasted that the USSR was "insuring" the energy future of Western Europe, a statement ill received within the EEC. One week later, reports emanating from Paris stated that a major explosion had occurred at a compressor station in the Urengoy field, causing a six-month delay in gas deliveries. That report was never substantiated, and it is now believed to have been either invented or greatly exaggerated by the French in retaliation for the Soviets' propaganda posturing.[61] By the end of 1985, most of the compressor stations were reportedly on line, and gas was flowing to western European homes and factories only a few months behind schedule.

7

A Legal Mess

Prime Minister Thatcher's strong opposition to the American sanctions was based mostly on what she believed to be the not only unfair, but also illegal, extraterritorial application of U.S. policy. For the prime minister, the sanctions constituted a threefold affront. First, they were retroactive; that is, they sought to void preexisting contracts. Second, they attempted to prohibit European companies from using legally acquired U.S.-origin technology for the pipeline. Third, they claimed jurisdiction over technology of European origin that the overseas subsidiaries of American companies had contracted to sell to the Soviets. Other European leaders shared Thatcher's view that Washington was meddling in the affairs of foreign companies and sovereign states over which it had no rightful control. This strongly felt common position cemented Europe's opposition to the pipeline embargo.

AMERICA'S EXTRATERRITORIAL CLAIM

Washington imposed its sanctions under the Export Administration Act of 1979. The Act allows the executive branch to control the export of American technology and products for (1) reasons of national security, usually involving the sale of military or other strategic goods to an unfriendly nation; (2) considerations of foreign policy whereby the controls would facilitate the achievement of foreign policy goals; and (3) cases where the export of the product in question would lead to a serious shortage of that product in the United States.[1] Initially, the

Reagan administration justified its application of the Export Administration Act as a tool of foreign policy to end the state of martial law in Poland. Later, administration officials invoked the national security clause on the pretext that the pipeline posed a strategic threat to the Atlantic Alliance.

The administration extended the sanctions to foreign subsidiaries of American companies and to foreign firms using American technology by arguing that these companies implicitly submitted themselves to the jurisdiction of American law, which purportedly applies to American-built or -licensed technology, and to companies based abroad that are American-controlled. Moreover, several of the contracts between American manufacturers and their European licensees explicitly recognized such jurisdiction. Alsthom-Atlantique's contract with General Electric recognized the applicability of American export controls:

> To facilitate the furnishing of data under this agreement, Alsthom hereby gives its assurance in regard to any data of GE origin that unless prior authorization is obtained from the U.S. Office of Export Administration, Alsthom will not knowingly . . . export to any country [in] group "Y" any direct produce of such technical data if such direct produce is identified by the code letter "A." Alsthom further undertakes to keep itself fully informed of the regulations, including amendments and changes thereto, and agrees to comply therewith.[2]

Group "Y" refers to a category of nations that includes the Soviet Union. Letter "A" is the status President Reagan assigned to the pipeline equipment when he extended the sanctions. According to this argument, Alsthom, and several other companies, were thus obligated to obey Washington's embargo. When they did not, they violated both the contracts they had signed with their American partners and U.S. law.

EUROPE'S REBUTTAL

In a note of protest to the Departments of State and Commerce in August 1982, the European Community challenged the retroactive nature of the sanctions. It made a compelling case. By applying ex post facto controls, Washington sought to void valid contracts and to prohibit the use of American technology that was perfectly legal—indeed anticipated—at the time of the original transaction. Such an initiative may be both necessary and legitimate when national security is clearly at stake, and when the purchaser of American technology explicitly accepts the United States' right to revoke the buyer's privilege to use that technology as it sees fit. But to attack the sanctity of a contract in order to achieve foreign policy goals is a dangerous exercise of power that seems likely to backfire, especially when the contract contains no revocation clause. Foreign trading partners will ultimately be pushed to look elsewhere if they conclude that American contracts are not worth the paper they are written on. Retroactive controls put a chill on the imposing country's ability to do business abroad.

The EEC also disputed the Export Administration Act and the Administration's application of the act. The statement affirmed that the American legislation could not be considered binding because it violated the principle of territoriality, which prohibits one state from unilaterally applying its laws on the territory of another. "The steps undertaken by the United States," the Europeans wrote, "are an obvious violation of this principle because they seek to regulate the activities of European companies that in no way fall under the territorial jurisdiction of America."[3]

The note went on to argue that the administration's use of the Export Administration Act did not adhere to the prerequisites set by Congress. According to the act, the president must: (1) take into account the reaction of other nations to the establishment or extension of export controls by the United States; (2) assure that sufficient efforts have been made to obtain the results desired first by negotiations or measures other than export controls; and (3) make every effort to engage and successfully conclude negotiations aimed at assuring the cooperation of foreign governments in the control of exports of products or technologies comparable to those being subject to control.[4]

The United States, in the European view, had violated all three of these requirements. The president, they claimed, did not sufficiently take into account European reaction, not did he attempt to reach a compromise on the pipeline issue. Instead, Washington took unilateral action in total disregard of the allies' position. Nor did the administration seek to obtain assurances from the Europeans that they would not export comparable technology to the Soviet Union.[5] One might add to this that the phrase in the second clause about "results desired" leaves Washington open to charges that it never made clear its objectives—did the United States want to end martial law in Poland, stop the pipeline, force the adoption of a tougher credit policy, or wage economic warfare?

Meanwhile, European companies found themselves in a Catch-22 situation. If they failed to comply with the American embargo, they would be subject to fines and blacklisting, a measure that would deny them access to American goods, services, and technology as well as to the American market. But once the European governments ordered the companies to fulfill their contractual obligations toward the Soviet Union, refusal to accede to their own governments' directives would have meant penalties payable both at home and—by virtue of the pipeline contracts—to the Soviets.

The debate in both government and legal circles sparked by the question of extraterritoriality continues today.[6] It cannot be resolved by weighing the strength of the arguments put forth. Each side is "correct" in its analysis. The cases they make, however, are mutually exclusive. As one expert on international law says, "Just because the United States has the authority to reach beyond its borders to insure compliance with its own laws, it does not necessarily follow that the other country cannot block the application of this transnational reach. Both concepts are valid, both in national law and international law."[7]

ONE SOLUTION

Perhaps the best criteria for the extraterritorial application of a nation's laws would include (a) the extent to which the export of the products or technology in question would harm national security, and (b) political common sense. Many legal scholars endorse the idea of a "protective principle" in order to justify extraterritorial sanctions. In testimony before the Committee on Foreign Relations concerning the pipeline, Stanley Marcuss, a lawyer specializing in American export administration law, explained that the theory of the protective principle is that

> A country may punish acts committed by aliens in foreign countries where those acts diminish the security, territorial integrity or political independence of that country. . . . In the context of international trade, only those transactions that genuinely and directly affect the security of the United States—as opposed to transactions that merely create inconveniences or even reverses for U.S. foreign policy—arguably provide a legitimate case for assertion of extraterritorial jurisdiction under the protective principle.[8]

He continued:

> I assume that the U.S. government has made detailed factual assessments of the likely impact on U.S. security of the construction of the Soviet-European pipeline. But if those dangers to U.S. security are merely general and speculative or indirect, then the protective principle may not provide firm ground on which to defend proposals to exercise extraterritorial jurisdiction in connection with that pipeline.

Since the pipeline crisis, lobbying from the business community has produced an important change in American law that reflects an implicit acknowledgment of the protective principle. The 1985 amendments to the Export Administration Act provide that valid contracts may not be voided solely to advance foreign policy goals. There must be a "breach of the peace" that poses a "serious and direct threat to the strategic interest of the United States." Moreover, the interruption of a contract must be "instrumental in remedying the situation posing the direct threat." Although the language "breach of the peace" and "serious and direct threat" is open to interpretation, congressional intent is clear: to protect contract sanctity, to restrict the president's power to use export controls as a tool of foreign policy, and to avoid more international legal muddles like the one that surrounded the pipeline.[9]

Political common sense may be the most important basis for deciding whether to invoke the extraterritorial jurisdiction of U.S. law. Another expert on export administration, Douglas Rosenthal, asked the Foreign Relations Committee if the United States

> . . .would stand for a similar application of foreign sovereign law applied to us? . . . Let us think for a minute about what kind of chaos there would be in international trade if other nations took a lesson from our book and applied their laws as aggressively against us.

Rosenthal then advanced this hypothetical example:

> Assume that the new French government became convinced that the military junta in Chile or some other country was an implacable enemy of their people. It is not an altogether implausible assumption. Assume further that some French companies had licensed important technology to certain unrelated U.S. companies years ago and that the French government was now threatening these U.S. companies with severe sanctions under French law if they honored contracts for the sale of

U.S.-made goods, manufactured in part based on this technology, to Chilean companies. . . . Can one seriously believe that this Administration or the majority of the American people would tolerate [such a] French dictate?

Rosenthal concluded:

The jurisdictional scope of the Export Administration Act must, if we are to avoid hypocrisy and a breakdown of our trading relationships and political alliances, be read as imposing limits consistent with the principle of comity, of reasonableness, of a proper balancing between conflicting national laws.[10]

Leaving aside the question of what the Export Administration Act should or should not permit, the issue of whether the administration adhered to the letter of the Act is equally ambiguous. The European Community's complaint that Washington failed to meet congressional requisites for applying the Act cuts both ways. Who is to say, for example, that the president did not adequately consider the reaction of the allied governments to his sanctions? Perhaps he did, and concluded that such considerations were outweighed by policy exigencies. Similarly, the administration could claim that it did indeed attempt to persuade the allies to abandon the Euro-Siberian through diplomatic negotiations, but was unsuccessful in its efforts. As for the third requirement, Washington believed that it had a monopoly on the banned technology, which meant that the question of allied cooperation in withholding similar products was irrelevant.

Washington can make a reasonable case that it did not violate the provisions of the Export Administration Act in a legal sense. However, its actions appear to have run counter to the spirit of the law. Hearings on the Act in 1979 indicate that the legislative intent was to ensure that export controls remain the exception, not the rule, of American trade policy.[11] The president may, in fact, have weighed all the considerations mandated by law. Yet his heavy-handed extension of export controls, coupled with his failure to offer a consistent justification for his actions, suggests that he did not accord these considerations the weight that the authors of the Export Administration Act sought to grant them.

At the very least, the pipeline dispute confirmed that the United States and its western European allies have very different approaches to coexistence with the East in general, and to trade relations specifical-ly. Until now, the alliance has managed to paper over this division and has survived the occasional crisis. It may be that inter-alliance disputes are both inevitable and, in the final analysis, manageable; after all, the pipeline crisis has come and gone, yet the alliance remains.

But the public quarrel over the Siberian project was singularly disruptive, and alliance "solidarity" suffered, at least temporarily, perhaps its greatest blow. "Solidarity" is a vague, intangible concept that probably tells us little about the true state of the allied relationship. Looking underneath the gloss of diplomacy, however, it is possible to trace the lasting impact that the Siberian pipeline controversy has had upon the Atlantic Alliance.

8

The Aftermath

In the wake of the pipeline embargo, two different assessments have emerged regarding its implications for the alliance. Certain officials of the American government, as well as European observers, feel that the embargo and the temporary imposition of sanctions against European companies constituted the catalysts needed to push the Europeans into rethinking their attitude toward East-West trade and into adopting stricter policies based on the American model. Most Europeans and members of the American business community assert that the embargo did long-term damage to American business interests by making foreign countries and companies more reluctant to rely on U.S. equipment and technology. Time will show which of these two contradictory ways of thinking is correct.

RETHINKING EAST-WEST TRADE

When President Reagan lifted the pipeline embargo in November 1982, he announced that the allies had reciprocally agreed to undertake major studies of East-West trade relations. Despite strong protests from the French that these two developments should not be linked, the studies began and Washington used them as a justification for its hard line position on the Euro-Siberian issue. W. Allen Wallis, under secretary of state for economic affairs, called the studies "quite productive." He told Congress that they had helped to bring about "an understanding and a realization of the security risks that can be incurred in trade, of which we either have been unaware, or have not been giving sufficient weight [to], or we had not been able to get together and work together [on]."[1]

The studies were made by the IEA, which evaluated Western energy security and examined alternatives to Soviet energy for western Europe; the OECD, which examined Western credit policies toward the USSR; NATO, which studied the impact of East-West trade on alliance security; and COCOM, which considered proposals to strengthen its controls on strategic technology and to add oil and gas equipment to its embargo list.[2] These studies were completed by the end of 1983, but were kept confidential. Some details, however, found their way into the press and academic literature, and government officials described the contents and conclusions of the studies in interviews.

The IEA produced a vague alliance compromise concerning dependence on individual sources of energy in general, and natural gas in particular. It noted that desirability of developing alternatives to Soviet gas—particularly Norway's Troll fields—but recognized that this would be more expensive than purchasing gas from the USSR. The United States wanted the IEA to recommend that no country depend on one gas supplier for more than 30 percent of its annual gas needs. This would have forced countries like Austria to take less Soviet gas, and would have prevented the European countries that fell just under that limit from obtaining more energy in the future or from building a proposed additional pipeline parallel to the Euro-Siberian.

The EEC countries disputed the 30 percent "safety ceiling" by pointing to their dual-firing ability and plans to expand storage capacity. From the European perspective, the burden of proving vulnerability fell on the Americans; the necessity for a 30 percent ceiling did not seem to them to be substantiated by convincing evidence. The ceiling, in any event, was becoming a moot question—the additional pipeline project had been put on hold for economic reasons, and the Europeans were in the process of renegotiating their contracts with Moscow so as to take less, not more, Soviet gas.

Washington had little desire to become embroiled in a repeat of the pipeline controversy. Nevertheless, the allies did pledge to avoid "undue dependence" on any single source of energy, as well as to conduct an annual review of Western reliance on Soviet natural gas.[3] As is so often the case with allied agreements on East-West trade, the operative terms were not clearly defined. Just what constitutes "undue" dependence was not specified. The IEA studies thus did little to advance the debate in the West on the issue of energy dependence or to furnish meaningful guidelines for the future.

A similar lack of substantive progress is evident in the OECD study. Technically, the revised OECD Consensus that places the Soviet Union in the "rich" country category—thus requiring the higher 12 percent minimum interest rate on Western government credit and loans—is included in the study. But the revision was initiated before the pipeline dispute and was finalized in principle while the crisis was in full swing. Since the embargo was lifted, the allies have made little headway on commercial bank credits, the quantity of both government and private credits, and the loan insurance provided by companies like Hermes, which Washington considers a subsidy.

At a meeting in May 1983, the allies endorsed a series of bland OECD suggestions advocating that East-West trade be "guided by the indications of the market" and requesting member states to "exercise financial prudence without granting preferential treatment" to the East bloc.[4] The language of this agreement was strikingly similar to the final communiqué at the Versailles summit, written as the pipeline crisis was about to explode. It is the type of agreement into which each signatory can read whatever it pleases. Since the OECD accord, European countries have advanced fewer credits to the East. However, officials cite economic, not political, reasons for this development. "East bloc debt from countries like Romania and Poland seemed to be too great a risk, and the eastern Europeans themselves were cutting down on requests in order to right their trade imbalances," says an official in the British Department of Trade.[5] It appears that in the area of credits, the alliance dispute did little to bring European policy into line with Washington's views.

The NATO members did not get much beyond an agreement to the effect that the security implications of East-West trade must be taken into consideration more systematically in the future—a fairly meaningless result. The United States lobbied to upgrade NATO's Economic Secretariat in order to emphasize the link between economic and military security. The Europeans vetoed this initiative, fearing that Washington was trying to mold NATO into a tool for economic warfare.

The United States may have had greater success in COCOM. To what extent this was a result of the pipeline sanctions is difficult to know, since the COCOM list was scheduled for review before the embargo began. Some new technologies, such as robotics, superminicomputers, sophisticated computer software, and telephone equipment, were added to the embargo list, while some widely available or outdated products, such as medium-power personal computers, were removed.[6] American officials said they were pleased by the "new spirit of cooperation" in COCOM, particularly regarding the enforcement of embargoes.[7]

However, the Europeans continued to balk at incorporating a broader definition of the term "strategic" for which the Americans had lobbied. They declined to create a separate military subcommittee to ensure the systematic inclusion of the military perspective in licensing decisions. They would not agree to a U.S. request to make COCOM a formal organization based on a treaty. In addition, they refused to expand the list of oil and gas technology subject to embargo in COCOM. In March 1984, the Reagan administration approved the sale of $40 million in submersible drilling pumps to the Soviet Union for oil exploration and production. An unidentified administration official told the *New York Times* that approval of the sale came after the United States failed to get COCOM to add the pumps to its embargo list in January. At a cabinet meeting, President Reagan decided that the United States would no longer ban the export of oil and gas equipment by American companies if it could not get the Europeans to follow suit.[8] To do so would have meant, as it often had in the past, rendering it possible for foreign manufacturers to supplant their U.S. competitors without accomplishing any U.S. objectives.

According to a British representative to COCOM, the United States tried hard to add the oil and gas technology to the embargo list. "They just didn't have the evidence necessary to convince us," he says. The official attributes the additions to the COCOM list in other areas to "better homework by the Americans. They made some strong cases."[9]

In sum, with the possible exception of its successes in COCOM, the United States seems thus far to have gained little in the way of policy concessions from the Europeans on East-West trade—certainly not enough to justify the alliance dispute over the pipeline. Washington's attempts to portray the sanctions as a tactic for eliciting more allied cooperation on trade matters is, at best, a gross rationalization. The East-West studies, while clearly desirable in their own right, were mostly a means for Washington to back out of the crisis gracefully and lift the pipeline embargo. As such, they should have been accepted in good faith by the Europeans, who could have swallowed their legitimate concern about issue linkage for the sake of alliance unity. Even this result proved unattainable in light of the French reaction to Reagan's announcement.

To trumpet Washington's sanctions as a victory for the administration because they coerced the allies into rethinking East-West trade is unjustified by the evidence. One has to ask whether similar results—marginal as they were—could not have been achieved without playing the pipeline card. Several European officials insist that their countries would have agreed to undertake the studies voluntarily. "Before the pipeline dispute even arose, we were prepared to look at issues of East-West trade through studies in the various alliance organizations," says Helmut Schmidt. "In fact, we had an informal agreement at the Versailles summit to do these studies just to calm down the Americans. You can't say the sanctions made us do something we were willing to do anyway."[10]

For now, the benefits to the alliance from the Euro-Siberian dispute are difficult to see. Perhaps the future will show that the crisis has made western Europe more reluctant to engage in trade with the East without weighing more carefully the strategic implications of such interaction, or consulting more closely with Washington. This thesis remains to be proved.

GIVING AMERICAN BUSINESS A BAD NAME

The most obvious effect of the pipeline sanctions on American business was the loss of contracts with the Soviet Union. The U.S. government estimated that controls on oil and gas equipment and technology sales to the USSR cost U.S. companies between $300 million and $600 million in exports. This conservative estimate failed to note the probability of long-term market loss.

One of the more egregious examples concerns Caterpillar, the pipe layer manufacturer. By the mid-1970s, Caterpillar had captured 85 percent of the Soviet market for such machinery, while its most important competitor, the Japanese firm Komatsu, had only 15 percent. Today, in the wake of President Carter's 1978 export prohibitions and the 1982 pipeline sanctions, the respective market shares of the two companies have been exactly reversed. In March 1983, Komatsu won a bid to sell the USSR 500 pipe layers for $210 million. The Soviets indicated that, following the pipeline embargo and given the continued existence of some American export controls affecting Caterpillar, they preferred to deal with the Japanese firm.[11]

A more subtle, and potentially much more damaging, result of the sanctions concerns West-West trade. In its note of protest to the United States, the European Community remarked that the extraterritorial extension of American sanctions could have a long-term negative impact on U.S. business interests generally:

> One inevitable consequence [of the sanctions] would be to call in question the usefulness of technological links between European and American firms, if contracts can be nullified at any time by decision of the U.S. Administration. Another consequence to be feared is that the claim of U.S. jurisdiction accompanying American investments will create a resistance abroad to the flow of U.S. investment.[12]

Although it is too soon to assess adequately the strength of this argument, there is some evidence to show its validity both in the specific domain of pipeline-related technology and in the broader realm of American products and know-how.

Several decades ago, European manufacturers decided to rely on the GE turbine, which was considered the best made and most efficient available. The quality of the GE product remains uncontested, but the Europeans are now pushing companies like Rolls Royce to devote more resources to producing a competitive European turbine in order to free themselves from dependence on the American technology. Horst Kerlen, vice-president of the German compressor manufacturer AEG-Kanis, puts it bluntly: "There is a doubt, a lack of trust, a feeling against the United States, that is the worst thing to come out of this affair. We have to be very cautious now about any new contracts that would bind us so totally to the US."[13]

An official in the British Department of Trade and Industry states that since the pipeline, "numerous" companies that produce oil and gas equipment, computers, and high technology more generally have expressed strong reservations about incorporating U.S. technology into their products. "I can think of at least one specific case recently," she adds, "in which one of our companies used Japanese technology instead of American precisely for fear of extraterritorial reach."[14]

European government officials have found the embargo to be excellent justification for restricting American participation in their economies. The former British minister for trade, Peter Rees, praised the Monopolies and Mergers Commission for refusing to approve the takeover of the British Davy Engineering Company by the American Enserch Corporation, partly on the grounds of the threat posed by the potential extraterritorial application of U.S. export controls. "How prescient the MMC was on this point [in 1981] has been amply borne out by recent events over the export of oil and gas equipments and technology to the Soviet Union," he told the Royal Institute for International Affairs.[15]

The British Trade Department has added an official warning regarding U.S. export controls to its manual on export licensing.

> Exporters should also be aware that the U.S. claim control over exports from other countries, including the U.K., where these are of U.S. origin, include components of U.S. origin, or were manufactured using U.S. origin technology. Although such U.S. regulations are not valid in U.K. law, the U.S. authorities commonly penalise foreign companies which do not comply, by denying them access to U.S. goods or technology in the future. Where a company has a presence in the U.S., legal action may lead to the imposition of fines and other penalties.[16]

American diplomats in Europe also claim to have noticed a new-found reluctance on the part of Europeans to rely on American technology. A U.S. official stationed in France says that

> When people see that the United States can and will restrict technology flow for political reasons, there is obviously a tendency to avoid becoming dependent on these processes. I've heard much talk in Paris since the pipeline that European firms are now willing to pay a premium to be independent of American suppliers.[17]

The reaction to the American sanctions in Europe strongly emphasizes the fact that efforts to curb East-West trade often do extensive damage to West-West trade. Indeed, lobbying efforts against the pipeline sanctions in Congress came in part from business groups that did little or no trade with the East, whether directly or through foreign intermediaries. Major American trade groups such as the National Association of Manufacturers (NAM) and the U.S. Chamber of Commerce, which criticized the pipeline embargo, expressed concern about the disruptive effect that export controls can have not simply on East-West trade but also on American foreign trade in general.[18]

In testimony before the Senate Foreign Relations Committee, NAM's vice-president for international trade noted that controls affect the reputation of U.S. companies as suppliers, even those having nothing to do with the immediate controls at issue:

> The semiconductor industry, for example, felt that the precedents being established in the recent controls jeopardized their cross-licensing agreements throughout the world. The issue being raised by the industry was whether the direct product of U.S. technology which was exported under a general license could be subject to future foreign policy controls after the technology transfer had taken place. The pipeline controls certainly seemed to assert such jurisdiction.[19]

The first round of American sanctions, in December 1981, had a direct impact upon U.S. business by banning the export of American technology, manufactured in the United States, for the pipeline. The second round, even though it targeted subsidiaries of American companies and foreign firms, had an equally negative impact on American industry and could, in the long run, prove damaging to the U.S. economy. Foreign buyers of American technology learned that the incorporation of U.S. parts or know-how into their products or manufacturing processes poses a dual risk. First, the American technology might suddenly become unavailable due to export prohibitions. Second, for the same reason, other countries might hesitate to purchase their finished products because the products contain American technology. Certainly, the imposition of export controls warned countries and companies the world over that the United States is not the most reliable trading partner.

Forty years after the creation of the Western alliance and four years after one of its most divisive internal disputes, the Americans and the Europeans continue to hold very different views on East-West trade. Until one side adopts the position of the other, or until a common policy emerges from the seemingly incompatible philosophies that now prevail, all the factors necessary for a repetition of the pipeline fiasco will remain present, awaiting only another detonator. The lesson of the pipeline crisis is the need for a shared allied approach on the issue of trade with the Soviet bloc. Thus, a final query: Which approach could find general acceptance within the West and simultaneously meet the strategic and economic concerns of the alliance countries?

9

The Great Debate

The Western alliance countries have tried three distinct strategic approaches in their effort to formulate a satisfactory policy for trade relations with the East. This in large part explains the frequent disarray of the West's trade policy. What follows is an overview of the different schools of thought on East-West trade, and an attempt to discover which one offers the soundest basis for a coherent alliance policy.

The first may be termed trade denial. It characterized American thinking during the first two decades after World War II and, to a lesser extent, the first Reagan administration. Proponents of this approach espouse a zero-sum theory of East-West economic relations: because commerce with the West is of value to the East, it should be curtailed even if it benefits the West. They also argue that the Soviet economy is on the brink of collapse, and could be pushed over the edge if denied the benefits of Western trade, thus, they hope, leading to the demise of the Soviet system. The transfer of any type of technology from West to East should therefore be stopped.

Linkage, the second school of thought, was dominant in West Germany during the 1950s and in the United States during Henry Kissinger's tenure as secretary of state. In this view, trade can be used both to reward the Soviets for good behavior in the international arena and to punish their transgressions through selective denial of commercial transactions.

Third, there is the pro-trade approach, a strategy common to most of the western European allies since World War II. If trade with the East yields an economic benefit to the West, it should be pursued, provided it does not harm Western security. The pro-traders hope that, in the long run, integrating the East bloc into the world economy will lead to a positive and peaceful evolution of the Communist system.

TRADE DENIAL

Testifying before the Senate Foreign Relations Committee in 1982, former chairman of the President's Council of Economic Advisers Herbert Stein argued: ''The Soviet Union is our enemy and trading with them is trading with the enemy. . . . The threat to us will be greater the greater are their resources.''[1]

Stein's view neatly sums up U.S. trade policy toward the East throughout much of the postwar period and is shared by several influential members of the Reagan administration. In essence, those who seek to deny trade to the East are concerned that commerce will help the Communist countries increase their military might, either directly or by easing the choice between guns and butter that, they assert, East bloc leaders are continuously forced to make. To assess the validity of this approach, it is first necessary to evaluate the impact of Western trade on the military sector, and more broadly the economy, of the Soviet Union.

During the 1930s and under lend-lease, Western technology made a significant contribution to the infrastructure of the Soviet economy, as a number of scholars, notably Anthony Sutton, have shown. Following the war, however, direct American trade with the East was minimal and European trade was constrained by policy directives from Washington. It was precisely during this period that the Soviet military achieved some of its most impressive advances. Without the benefit of direct Western assistance, the Soviets were able to produce atomic and hydrogen bombs a few years after the Americans. They then stunned the world with Sputnik and the first intercontinental ballistic missile. Economic growth rates throughout the 1950s were among the highest in Soviet history.

It is possible that these advances reflect the long time needed to absorb the Western technology acquired before and during the war. Most scholars, however, believe Soviet military and economic progress in the 1950s and early 1960s demonstrates the USSR's ability to forgo Western technology. Some even argue that the quasi embargo on trade by Washington helped Stalin carry out his brutal internal policies by providing a scapegoat—the West—for the economic woes the country experienced.[2] While Western technology could have accelerated existing Soviet growth rates, and perhaps even have led to the development of other advanced weapons, it is clear that trade denial did not prevent noteworthy military and economic development in the Soviet Union. The historical justification for the success of such a policy simply does not exist.

Some proponents of trade denial take a short-term historical view and point to the substantial growth of the Soviet military throughout the 1970s, a time when East-West trade reached its peak. Increased trade apparently did nothing to check Soviet military expansion, and may have promoted it. There are two problems with this argument. First, significant debate exists over the extent of the Soviet military buildup. For example, many experts who based their claims of substantial Soviet military growth on the fact of increased defense expenditures failed to take into account a dramatic increase in the cost of producing military equipment in the USSR. Second, by 1978, trade with the West amounted to only 1.4 percent of Soviet GNP. The share of Western machinery in total Soviet equipment investment stood at under 5 percent. It is hard to justify that such limited trade substantially affected military growth, or even the general development of the Soviet economy.[3]

Equally difficult to prove is the contention that increasing Soviet export potential significantly benefits the military. The connection between Soviet hard currency earnings and military expenditures is impossible to ascertain with any certainty. An official of the CIA's National Intelligence Council, in a memorandum to the State Department shortly after the pipeline crisis, concluded:

> We continue to be highly skeptical of the usefulness of trying to justify [export] controls . . . using as a criterion primarily the special importance of Soviet hard currency earnings. It is extremely difficult to establish any direct link between Soviet hard currency earnings and military and foreign policy expenditures.[4]

Even if a direct link between East-West trade and Soviet military expenditures could be established, it does not follow that trade denial will lead to fewer Russian guns. Moscow has shown itself unwilling to sacrifice what it considers to be a necessary level of military spending for the sake of Soviet consumers, despite great economic difficulty. Thus, the annual growth rate of the Soviet economy faltered appreciably throughout the 1970s while defense spending increased. Between 1975 and 1980, the average economic growth rate was 2.8 percent,[5] while military spending over the same period increased at an annual average of 4.5 percent.[6] The rigid central control of the Soviet economy is such that Moscow can redirect resources to the military when it sees fit and tighten the belt elsewhere. Moreover, the Soviets have been reluctant to rely on the outside world for military technology in order to avoid dependence on an adversary.[7]

This is not to say that Western technology has no beneficial impact on the Soviet military. Helping Soviet economic performance results in savings in research costs and other gains from increased efficiency that are passed on to the military. But while selling the Soviets nonstrategic technology and products may result in indirect gains for the military, it does not follow that withholding such trade will produce a corresponding decrease in Soviet military might. The error is in assuming that a significant choice between guns and butter does exist from Moscow's perspective. Military spending has never been sacrificed in the past. Western trade simply makes the military burden a little easier for the Soviet citizens to bear. Most important, it has never been demonstrated that the USSR has obtained technology from the West that it could not have developed itself, given time and incentive.[8]

It is also possible that the West can actually derive a limited strategic advantage from selling high technology to the East. Some economists believe that a country selling the most advanced technology is also selling obsolescence. The vendor makes the purchaser dependent on it for the product in question because the purchaser neglects to develop indigenous research, development, or production. In the time it takes the purchaser to integrate the new technology into its economy, the vendor will have refined the product one step further, thus perpetuating dependence and creating a strategic advantage. The problem is further compounded for a buyer like the Soviet Union, which acquires much Western technology illegally. It must rely on expensive, risky, and highly uncertain tactics to assure itself state-of-the-art technology.

This theory does not, in practice, always hold up under scrutiny. The Japanese imported Western technology after World War II, then not only integrated it into their economy but also perfected it further, thus allowing them to win a substantial share of the world market for high technology. However, the Soviets have so far failed to show similar capabilities. With a few exceptions, the USSR has been unable to improve the products it imports, and is also poor at copying these products by "reverse engineering." While the Soviets do integrate Western technology into their system rapidly, once it is installed, it tends to perform with substantially less efficiency than it would in the West, because the archaic Soviet industrial base is ill equipped to handle modern technology. Soviet specialist Philip Hanson has estimated that Western technology in the USSR operates at only 60 percent efficiency.[9]

The most noteworthy example of the dependence trap involves the Soviet computer industry. In the early 1960s, Moscow decided to copy Western technology instead of relying on its own scientists to design and develop computers. When IBM in 1964 began producing the 360 series mainframes, the first third-generation computers manufactured in large volume, the Soviets set out to emulate them. They forced some of the East bloc satellites to join the effort and to abandon their indigenous computer development programs. Yet it was not until eight years and many production problems later that the USSR introduced its version of the 360—called the Ryad-1—by which time IBM was producing the more advanced 370 series. Another six years passed before the USSR developed the Ryad-2, based on the then outdated 370 series. Today, the Soviets lag eight to ten years behind the United States and Japan in the computer sector.[10]

Since World War II, numerous Soviet officials have expressed concern about relying on Western technology. Shortly before his death, Leonid Brezhnev told the newspaper *Pravda*: "We must examine why we sometimes forget our priorities and spend large sums of money to purchase equipment and technology from foreign countries that we are fully capable of producing ourselves, and often at higher quality." Similarly, A. P. Aleksandrov, president of the Soviet Academy of Sciences, said: "We must actively develop our national techniques and technology. We must not create gaps in our technological development by unjustifiably relying too extensively on foreign techniques."[11] Since Gorbachev's rise to power, a number of prominent figures, including the general secretary's scientific adviser and Boris Yeltsine, the party secretary for Moscow, have criticized the USSR's economic dependence on the West. Prime Minister Rijkov even took to the podium of the 27th Party Congress to call for a reevaluation of Soviet imports of Western technology. These statements could be evidence that while technology transfer allows the Soviets to take one step forward by skipping the research and development stage of technological innovation, it pushes them two steps backward by creating a reliance on Western know-how.

Recently, proponents of economic warfare have advanced a somewhat different argument. According to many Soviet specialists in the West, the Soviet economic system has reached a crisis due to its continued reliance upon the Stalinist economic model.[12] A drastic reform of the system is needed, but Soviet leaders fear the change will bring about wide-ranging disruptions in the political and social systems as well. Proponents of trade denial say that Western technology has allowed the Soviets to delay making such changes, thereby helping to maintain the political and social status quo. Political scientist Samuel Huntington restated this view in 1978:

> Confronted with a tightening labor supply, declining factor productivity, lower rates of investment, continued agricultural problems and uncertainties, and a resulting decline in the overall growth rate, the Soviets could have responded broadly in three ways: by a return to a mobilization economy, involving autarky and repression; by fundamental reforms in the organization and incentive structure of the Soviet economy, which would have had unsettling and potentially threatening political consequences; or by intensified economic engagement with the West, which would compensate for Soviet economic shortcomings by massive imports of Western equipment and technology. The third option provided them with the least costly way of avoiding economic reforms, continuing a substantial military build-up . . . maintaining a minimally acceptable rate of economic growth, and securing the critical technology needed in certain key industrial areas.[13]

It follows that U.S. policy should seek to deny the Soviets Western technology and thereby leave Moscow no choice other than a dramatic reform of the system. Leading members of the Reagan administration have espoused just such a policy. Former National Security Adviser William Clark, for example, argued that Washington should "force our principal adversary, the Soviet Union, to bear the brunt of its economic shortcomings."[14]

While this proposal has a certain appeal, it raises further questions. For example, is the state of the Soviet economy truly so precarious that an embargo on the sale of Western technology would necessitate a complete overhaul? Marshall Goldman, codirector of Harvard's Russian Research Center, who has written extensively on that system's failures, concludes that, given the USSR's tremendous economic base and resources, collapse is far from imminent.[15] A range of analysts, from the CIA to the Wharton Econometric Forecasting Associates, echo Goldman. Henry Rowen, chairman of the CIA's National Intelligence

Council, told the Joint Economic Committee of Congress that, despite important shortcomings, there is "not even a remote possibility" that the Soviet economy will collapse, adding:

> The ability of the Soviet economy to remain viable in the absence of imports is much greater than that of most, possibly all, other industrialized economies. Consequently, the susceptibility of the Soviet Union to [economic denial] tends to be limited.[16]

Despite such assessments, Washington has made a modest and uncoordinated attempt at trade denial since the invasion of Afghanistan. Although overall U.S. exports to the USSR total more than $1 billion a year, at least 70 percent of this is in agricultural trade.[17] Nonagricultural exports dropped from $819 million in 1976 to under $500 million in 1984. Still, the Soviet economy continues its course.

It is true, however, that Soviet leader Gorbachev has made economic reform his first priority. Several of the men he promoted to influential posts are from the Georgian Republic, renowned for its economic innovations—including some that compromise Soviet ideology. For example, as first secretary of the Georgian Communist Party, foreign minister Eduard Shevardnadze introduced a number of "market-oriented" reforms, such as bonuses for farmers who surpassed their production quotas and greater managerial independence in industry. The reforms have proved to be highly successful, as evidenced by significant increases in the output of fruit, vegetables, and even wheat. Between 1981 and 1985, industrial production in Georgia rose 33 percent.[18] Nevertheless, a revolution in the structure of the Soviet economy does not appear to be likely. Soviet leaders, including the new generation that favors innovation, are worried about the potential impact of reform on the social and political systems generally. They also must deal with domestic political restraints in the form of antipathy to change from the remaining, and still powerful, members of the pre-Gorbachev bureaucracy. Soviet economic reform, if it is to succeed at all, seems destined to be evolutionary, not revolutionary, in nature.

Even if the Soviet economy were in the dire straits some claim, it is highly questionable whether curtailing commerce would promote reform, in view of the relative insignificance of Western trade to the Soviet economy. The modest reforms now being implemented simply demonstrate Soviet awareness of the deficiencies of their system.

Assuming that halting East-West trade would in fact lead to the collapse of the Soviet economy, would the West escape unscathed from such an upheaval? Perhaps. More likely, Moscow would blame the United States and its allies for Soviet economic failures, using the embargo as a reason for more repressive action at home and more aggressive action abroad. While Communist ideology may be bankrupt, Russian nationalism can still claim millions of adherents among the populace. The proven reflex of the Russian people to rally around the flag in times of difficulty, even when their leader is someone as despicable as Stalin, makes withholding most trade fruitless. Waging aggressive economic battle—for example, by attempting to spend the Soviets into bankruptcy through a new arms race—might be even worse, given the resulting waste of our own resources. Successful economic warfare could prove much more dangerous to the West than the sale of peaceful, nonstrategic technology to Moscow.[19]

Whatever the drawbacks of trade denial, and despite the quantitative limits of East-West trade, it is nonetheless true that the Eastern bloc often acquires technology that should remain in friendly hands. Some of this is "dual use" technology, the term applied to products with both civilian and military applications. In the 1970s, for example, the West helped the Soviets build the massive Kama River truck plant; a few years later, vehicles produced at that plant were used to transport troops in Afghanistan. Computer technology provides a compelling dual use dilemma. A computer purchased to allocate resources in a clothing factory can also coordinate a city's defense plan.

Washington has complained that lax European export controls allow too much dual use technology to fall into Soviet hands. Many Europeans allege that Washington labels everything sold to the East as capable of dual use as a pretext to impose a trade embargo, and not to identify technology truly critical to the Soviets. This has occasionally paralyzed efforts by the alliance to coordinate export controls.

In COCOM, American and European representatives consistently disagree on a definition of that organization's key operative term, "strategic." The Europeans view the word as covering items with military applications only, while the Americans want a broader definition that would include goods that contribute even indirectly to Soviet military strength.[20] The debate intensified from 1976 on, when an American government-sponsored review of export controls (the Bucy

Report) urged that the West redefine its definition of strategic technology to place greater emphasis on knowledge, skill, and information as opposed to tangible items. While the report advocated a decrease in the control of end products, its focus on "critical technologies" that "transfer vital design and manufacturing know-how most effectively" raised concern both in Europe and the United States that the free flow of information within the West could be impeded. Also, the Europeans correctly read the report as an indictment of East-West projects involving turnkey factories, joint ventures, technical exchange agreements, and training programs. To them, such a critique would facilitate efforts to curb peaceful, nonstrategic East-West trade.[21]

The United States has frequently tried to place numerous technologies and products on the multilateral embargo list despite staunch allied opposition. The Export Administration Act mandated the creation of a Militarily Critical Technologies List (MCTL) that, it was hoped, COCOM would adopt. However, the document, which was classified until 1984, is so extensive in its coverage that it alienated the Europeans rather than promoting their cooperation. According to testimony before the U.S. Senate, the list is "700 pages long . . . and includes virtually all modern industrial technology."[22]

European qualms in COCOM result in large measure from their belief that once the definition of "strategic" is extended to include items without direct military applications, the extension will continue infinitely and curtail "innocent" trade. The ultimate logic of the position that nonmilitary trade is strategic is that *all* trade helps the Soviet Union and should be stopped. In this regard, the proponents of free trade like to cite an anecdote attributed to Khrushchev. The former Soviet leader supposedly said that the most strategic product America and her allies should seek to deny Russia is the common button. Why? Because if the Red Army troops had buttons to hold up their pants, they could aim their rifles with two hands instead of one.[23] The story makes a point: Whenever American officials criticize the Europeans for exporting high technology to the East, the latter invariably respond that American grain is an even more strategic commodity because it feeds the Red Army.[24]

The Europeans refuse to allow COCOM to be transformed into a political instrument for coordinating American-imposed trade sanctions. They also balk at U.S. attempts to hide foreign policy objectives behind national security rhetoric. The politicization of COCOM only leads to that organization's paralysis and prevents the achievement of legitimate U.S. goals. For example, when Washington pushed the Europeans to broaden the definition of "strategic" and refused European requests to remove some outdated, widely available technology from the embargo list, discussions over updating the computer list broke down. As a result, COCOM relied on an early 1970s computer embargo list until 1984. Given the rapid evolution of computer technology, the effect was doubly negative: entire new generations of computers were theoretically not subject to COCOM control while outdated hardware remained embargoed.

While the sale of dual use technology irritates Washington, there is general agreement among government and academic experts in the United States that such "legitimate" trade provides little truly critical technology to the East. To the extent that a problem exists, it is not dual use exports but the illegal transfer of technology. A 1985 Department of Defense report claimed that 5,000 Soviet military research projects each year benefit significantly from Western-acquired technology. At least one-fourth of these items are on export prohibition lists, the report said. The CIA estimates that 70 percent of Soviet military gains from Western technology are the result of theft and industrial espionage.[25]

Washington's attempts to resolve this problem, however, have run into major stumbling blocks, some of its own making. Instead of seeking to enforce existing prohibitions more efficiently, the administration has tried to expand export controls. Besides being impractical, this policy has alienated the Europeans and the American business community. As a result, illegal technology transfer, which can be reduced only through cooperation, remains both a strategic problem and a political issue within the alliance.

For example, when the Commerce Department in 1984 proposed tighter licensing requirements on products to be sold abroad that it feared might be reexported to the East bloc, the business community offered what one Commerce official called "a record number of negative comments."[26] American companies were especially upset by two provisions in the proposed regulations. One rule required certain foreign recipients of U.S. technology to provide a list of the names and addresses of all their potential customers who might receive that technology through one of the recipient's finished products. A second provision curbed distribution licenses, which allow American exporters to make multiple shipments over an extended period of time under a single license instead of requiring them to apply for a new one for each individual transaction.[27]

Both provisions, exporters complained, would create long delays in obtaining an export license and, by slowing business, would decrease American competitiveness. The "list" rule meant, for example, that the Commerce Department would have to review at least 1 million names—200,000 from IBM customers alone—that the foreign recipients of U.S. technology would be forced to provide.

The curb on distribution licenses was an even greater issue. According to the NAM, at least $20 billion a year in American products is exported under distribution licenses. Dramatically decreasing their number would cost American companies millions of dollars in additional paperwork and perhaps billions in lost revenue, because foreign companies using American technology and foreign distributors of American technology would prefer to buy similar products from non-American sources in order to avoid processing delays.

Some major U.S. multinational companies provided examples of losses that would result from the new regulations. Data General estimated that 90 percent of export revenues for the computer and semiconducter industries flow from distribution licenses; limiting the licenses could mean the loss of 40 percent of those revenues. For Data General alone, this would translate into a loss of $1 billion in revenue and approximately 20,000 jobs over five years. IBM, which derives 43 percent of its worldwide gross income from sources outside the United States, noted that over 70 percent of its foreign sales relied on distribution licenses. The Commerce Department predicted that it would have to issue approximately 1 million individual licenses every year instead of the current 110,000 distribution licenses if the regulations entered into effect. The time needed to obtain most licenses was expected to increase from less than a month to between 60 and 90 days.[28]

American companies also criticized the Commerce Department for placing more products on the East bloc embargo list. They argued that most of the newly banned items should not be considered "critical technologies." Moreover, they asserted that it was pointless to embargo some of the products that did, in fact, fall into that category, because they were openly available to the East in other countries. Even if Washington managed to win COCOM approval for an expanded embargo list, non-COCOM high technology producers such as Switzerland, Sweden, Finland, Australia, South Korea, and India might fill the gap.

Business leaders and legislators from high technology states lobbied the administration to modify the regulations. Massachusetts Governor Michael Dukakis, in a letter to President Reagan, wrote: "A conservative estimate is that if the regulations are promulgated, 20,000 jobs, representing 10 percent of all jobs in high technology in Massachusetts, will be lost within one year. Companies in our state alone will lose approximately $2 billion in sales."[29] As a result of these efforts, a compromise providing more self-policing by American companies was adopted.

The debate over the Commerce Department proposals reveals a fallacy in the case made by proponents of economic warfare against East-West trade. They argue that, because commerce with the East is of "trivial economic consequence to the U.S.,"[30] it can be used as a weapon of American foreign policy without damaging America's economic interests. But it is clear that to curb East-West trade effectively, West-West trade also must suffer. At a time when the trade deficit amounts to more than $150 billion a year, that presents a particularly unattractive prospect.

The dangers to the American economy and, ultimately, to national security in restricting West-West trade in high technology are evident. The high technology sector, one of the most dynamic parts of the economy, exports 30 percent of its output, compared with 8 percent for U.S. manufacturers as a whole.[31] These exports are necessary if the companies are to prosper and continue to innovate through research and development. Unnecessary export controls place U.S. firms at a competitive disadvantage in sales to non-Communist countries and stigmatize them the world over as unreliable suppliers and trading partners. Why should a Swedish firm, for example, buy a computer chip from an American company requiring at least 60 days to obtain the necessary export license, and run the additional risk that the American partner might have to cancel the contract if Washington so decides, when the same product is available in Japan with no licensing delays and little threat of subsequent cancellations?

High tech manufacturers are also essential to the U.S. defense. They supply over two-thirds of all the hardware purchased by the American military.[32] Here again, curtailing exports and thereby eliminating the advantages of scale production means hindering research and development for the military.

Conflict between the need to control strategic exports and to allow U.S. manufacturers the freedom they require to do business abroad is inevitable. The problem has intensified since the mid-1970s because of changes in the way technology is developed in the United States. Until the 1970s, the bulk of research and development for technologically sophisticated goods was undertaken by the government or under government contract for use in the space and military programs. Today, most advanced technology is first developed in the private sector, often by small, young companies burdened by high start-up costs and slow market penetration. If a military application is found for their product and the government feels compelled to impose export controls, the economic results are self-evident. For this reason, there is much to be said for limiting such government intervention to cases where it is truly necessary for national security.

Nearly 40 percent of all manufactured goods require some type of license for export. Many of these products are low-end, not cutting edge technology, or they are widely available in other countries. This leads to much unnecessary, ineffective regulation that hinders American industry without advancing national security. Until the 1985 amendments to the Export Administration Act, a thermostat that anyone could buy in a hardware store required a distribution license for export because it contained an imbedded microprocessor that the Defense Department considered strategically valuable to unfriendly countries. Even certain microwave ovens could only be sold under an export license.

Despite the 1985 amendments, hundreds of examples of useless restrictions remain. The manufacturers of sophisticated hand-held calculators for sale in any electronics shop need a license to export to non-COCOM countries, even allies such as Sweden and Australia. So do the makers of the machines that descramble satellite signals for television—items that someone with $20 and a high school diploma could make him- or herself. The Commerce Department requires a license to sell the filament used in a common light bulb to any country except Canada. Similarly, a company wishing to export a 250 megahertz oscilloscope—a machine such as a heart monitor that converts signals, impulses, or sound waves into a visual image—must have a license, despite the fact that the Soviets and the Chinese, not to mention the French, Dutch, and Japanese, make oscilloscopes with a superior 300 to 350 megahertz capacity.[33]

While license applications for most of these products are almost always approved, the U.S. manufacturer must incur substantial licensing costs and delays that reduce its competitive advantage. President

Reagan's Commission on Industrial Competitiveness, headed by Hewlett-Packard's chief executive officer John A. Young, in 1983 estimated the annual lost sales as a result of control policy at $12 billion. Howard Lewis, the NAM's vice-president for international trade, argues: "Either it's absurd to let sophisticated technology like that be produced and sold so easily, or it's crazy to control products that can be found in every American living room."[34]

To be fair, Washington has tried to better enforce existing export controls, and not simply expand them. The difficulties it has encountered prove the practical complexity of that task.

In 1981, the administration established "Operation Exodus," a program that augmented the Customs Service's budget by $30 million and increased the number of its field agents from 7 to 29.[35] Exodus produced several "espionage coups," most notably the seizure of a sophisticated Vax super minicomputer in West Germany just hours before it was to be shipped illegally to the USSR. At a press conference attended by both the secretary of defense and the secretary of the treasury, administration officials expressed their pleasure at having prevented the Soviets from acquiring a machine with numerous potential military applications. But the officials neglected to mention that the Soviets almost certainly had already acquired close to 20 Vaxs. Assistant Secretary of Defense Richard Perle told the *Wall Street Journal* that "Dozens of Vaxs are missing, and it's likely the Soviets have them."[36]

There is a great deal of additional evidence to show that Exodus and other government efforts have done little to stop the Soviets from acquiring high technology in the West. At the first International Machine Tool Exhibition in Moscow during the spring of 1984, American visitors were amazed to find state-of-the-art, computer-controlled machines using Western technology on display. Much of the technology was on the American embargo list. "What we saw was an eye-opener," said James Gray, president of the National Machine Tool Builders Association. "Only the most naive person would believe that we are preventing the Soviets from getting technology. We don't control nearly as much technology as we think we do."[37]

As American efforts to halt the flow of technology eastward increase, so does the Soviet procurement effort. Intelligence officials in Washington estimate that as much as 90 percent of the East bloc diplomatic personnel in this country concentrate on acquiring U.S. technology and industrial secrets. The Soviets organize or utilize companies in western European countries to reexport technology to the

USSR. They also use "fronts" to buy and study the equipment without assuming the risk of shipping it to Moscow. Some items that Washington would like to control, such as sophisticated computer programs, are so small that they can be smuggled in a briefcase or a diplomatic pouch.

Most important of all, many officials agree that there are simply too many products subject to control to make effective policing possible. Nearly 300,000 different products now require a special license for export. "If the ultimate objective is to stop the illegal flow, we are never going to achieve that," says Acting Assistant Secretary of Commerce William Archey. "There are just too many things to watch and too many ways to get them out of an open society."[38] If Washington's goal is simply to hinder the Soviet procurement effort, it must not forget that achieving even that limited objective imposes costs on American businesses and the U.S. economy.

What is to be done? We could, as the Reagan administration has suggested, limit foreign scholars' access to universities, prohibit the free circulation of technical papers, and remove products from the national and international marketplace. We could attempt to persuade the Europeans to do the same. But the cost, in terms of progress, economic efficiency, and, ultimately, freedom would simply be too high. It could far exceed whatever the Soviets gain from access to our technology. In effect, we are adopting an R&D policy from the Soviet bloc that is fundamentally hostile to innovation. Far from selling the Soviets the rope, we are importing it from the East. More fundamentally, Western disgust with totalitarianism centers on the closed society that system produces. By placing unreasonable restrictions on exports and on the development of ideas, we move in the direction of our adversary.

The point is not that strategic export controls are not necessary. They are. But controls that do not work are worse than no controls at all. The administration must balance the need for export controls against American business interests, the politics of prohibitions, and the technical barriers to embargoes. This balancing reveals that American interests would be best served by streamlining the embargo list so that it covers only technology whose primary application is military. Rather than build short fences around broad areas, we should erect tall barriers around carefully defined areas. Such an initiative would go far in eliciting more cooperation from the western Europeans and American business, and make it easier to control those products that Washington concludes must be embargoed.

LINKAGE

Under the Nixon and Ford administrations, American trade policy toward the East attempted to link commerce with Soviet behavior. This was similar to the approach that West Germany adopted under Chancellor Adenauer in the 1950s. Unlike economic warfare, the linkage theory requires increased economic ties with the East, which are, in turn, used as both a carrot and a stick to influence the conduct of East bloc nations. Henry Kissinger writes: "Our strategy was to use trade concessions as a political instrument, withholding them when Soviet conduct was adventurous, and granting them in measured doses when the Soviets behaved cooperatively."[39] Thus, while economic warfare seeks a halt to nearly all trade between East and West, linkage employs trade denial only in response to a specific, undesired action by an adversary.

From its inception, Kissinger's strategy encountered obstacles and exhibited flaws that rendered it less potent a weapon than its proponents had hoped. The stumbling blocks were not of Kissinger's making; whereas the secretary had been clear that linkage should be applied only to mold Soviet conduct abroad, Congress, and later President Carter, attempted to use trade to dictate Soviet domestic policies. As we have seen, Congress, through the Jackson/Vanik amendment, tied the 1972 trade agreement to emigration levels, and Carter imposed an embargo on technology exports to the Soviets following the arrest of two prominent dissidents. Moscow, however, refused to submit to pressure aimed at what it viewed as its domestic affairs, particularly when the complaints were aired in the international media. Although the Soviets have shown some willingness to link human rights concessions informally to trade, as soon as the connection is made explicit, and public, the Soviets balk.

To be successful, linkage has certain requirements that are not easily met. Using the stick by way of "selective" trade denial necessitates that the wielder be in a position of leverage. During the 1950s, the West Germans discovered that their attempts at linkage usually failed because German trade lacked decisive importance to the Soviet economy. Furthermore, despite the increase in East-West trade during the détente years, it never represented more than a modest portion of Soviet GNP. Effective use of the stick depended on the construction of Kissinger's "network of relationships," which was never given a chance to evolve.

In the short term, the threat of lost opportunity and the temptation of the carrot were thus the only tools at Kissinger's disposal. With these, Washington achieved some success. Nixon has argued that the Soviets would not have agreed to the first Strategic Arms Limitation Treaty in 1972 without the prospect of increased trade with the United States: "The Soviet leader's eagerness for trade was one of our most powerful levers in any concessions on political issues. . . . We might not have signed SALT I otherwise."[40] Similarly, the Soviets at one point had agreed to permit more emigration in return for MFN status, repudiating the agreement only after it had been made public.

Since 1985, there has been talk in Washington of a return to "positive" linkage. Business groups and some leaders of the American Jewish community have urged Congress to consider repealing the Jackson/Vanik amendment as an incentive for the Soviets to allow more emigration. Moscow has dropped strong hints that it would agree to such a trade-off, from which both sides would likely gain. Jewish emigration was cut by two-thirds following the passage of Jackson/Vanik—from 33,500 visas issued in 1973 to 13,000 issued in 1975. Today, the number of Jews allowed to leave every year is in the hundreds.

Nonetheless, while a quiet deal is desirable, there should be no illusions about the scope of the resulting benefits. The Soviets surely find it tempting to create a more favorable trading and diplomatic climate by allowing more Jews to leave. Increased emigration—while an unwanted admission of the unattractiveness of Soviet life—is a concession to the West that involves little real cost to the USSR. Yet it seems unlikely that the Soviets would abandon the pursuit of domestic or foreign policy objectives they considered important to their national interest for the sake of *potential* economic gains, especially when such gains promise to be very modest. Even if Jackson/Vanik is repealed, substantial restrictions on credits and exports to the East will remain. Unless these, too, are liberalized, more visas for Jews and other "refuseniks" are probably the most progress the West can hope to achieve by feeding Moscow the Jackson/Vanik carrot.

Even in the long run, a policy of linkage may have proved to be problematic. The network of trade relationships it requires is a two-way street that produces interdependence, not simply dependence. Using "stick" linkage can be not simply inneffective, but also counterproductive. Carter's flawed grain embargo makes the point, just as Reagan's pipeline embargo did a few years later.

In a study of the grain embargo, Robert Paarlberg writes that sanctions, if they are to be effective, require "an unbroken chain of favorable developments, in three distinct areas, all at the same time."[41] These developments are control by the country imposing the embargo on the product in question within its own boundaries; control by the embargo-imposing country over other exporters of the product; and sufficient leverage by the embargo-imposing country on the target country. Should any of these links be weak, the entire strategy can fail.

As we have seen, the third prerequisite is often lacking in East-West economic relations where dependence by either bloc on the other is low; however, in specific instances, it can exist. In the case of the grain embargo, American officials knew that the Soviets relied to a large extent on external sources—particularly the United States—for their supply of grain. At this point, though, the other links of the chain were weak and made the embargo relatively inneffective. First, the United States could not prevent other countries, including its western European, Canadian, Australian, and Argentine allies, from selling grain to the Soviets to replace the denied American supplies. Moscow quickly managed to make up most, though not all, of the embargoed grain, and American leverage was minimized. Second, control over domestic grain producers gradually weakened as the farm lobby pressured Carter, and then Reagan, to rescind the embargo. Paarlberg's chain was broken.

While the embargo may not have been the total failure its critics allege, since it did in fact produce grain shortages in the Soviet Union, it was not a success by most measures. The stated goal of the embargo was to force the Soviets to leave Afghanistan. That did not happen. Nor did a panic slaughter of livestock in the USSR materialize, for Moscow imported record quantities of meat from abroad to offset the decline in domestic meat production. Moreover, Carter's initiative imposed costs on the West. American farmers lost a substantial portion of the Soviet market, their share decreasing from 70 percent to 34 percent. And while U.S. grain producers now enjoy a larger share of the world market than they did prior to the embargo, one must recall that the Carter administration had to provide a $3 billion subsidy program to offset the short-term losses that followed the anti-Soviet embargo.[42]

The relative failure of the grain embargo is simply another example of the generally sorry history of economic sanctions during this century. The sanctions voted by the League of Nations against Italy did nothing to prevent Mussolini's aggression in Ethiopia. Franco's Spain survived a coordinated embargo by the Western democracies, and Castro's Cuba surmounted an American embargo. Soviet attempts to hold satellites such as Yugoslavia, Romania, and China in its orbit with an economic leash failed. The Arab nations' boycott of Israel has also proved ineffective. An exhaustive study by Gary Hufbauer and Jeffrey Schott of the Institute for International Economics concluded: "In most cases sanctions do not contribute very much to the achievement of foreign policy goals. . . . Sanctions are a decreasingly useful policy instrument."[43] This is especially true in the area of East-West trade. To succeed, sanctions must be coordinated with our allies. Yet, given the disagreements within the West over the desirability of even limited trade denial, especially when such denial seeks to further one nation's foreign policy strategy as opposed to its presumably more compelling national security goals, cooperation is elusive.

This does not mean that sanctions are useless. Even when Paarlberg's chain is weak, sanctions are a viable policy option if the imposing country wants to make clear displeasure, either to its allies or to its adversaries, in response to actions it deems unacceptable. They may even be necessary if the other options are less effective or more dangerous. Also, such sanctions can have a beneficial secondary effect. In the case of the grain embargo, even though the American ban did not succeed in forcing the Soviets to withdraw from Afghanistan, it may have made similar forays less likely in the future by imposing even limited, short-term costs. Some scholars have even claimed that the embargo was in the process of forcing a major reform of the Soviet agricultural system when it was lifted by Reagan.[44]

But while sanctions may have their place in a nation's stock of policy tools, the fact that they are also awkward, often ineffective, bad for the embargoing nation's reputation as a reliable supplier, and costly to that country makes stick linkage a questionable policy. Ironically, targeted embargoes may be more easily employed by totalitarian regimes than by free-market economies. As one economist writes:

> A market economy is not as well placed as a command system to absorb and diffuse the shock of withheld trade sustained in the pursuit of political objectives. Nor is it well placed to deliver the shock. . . . Except under wartime conditions [when the short-term benefits of an embargo could prove decisive] market oriented nations cannot expect to employ sanctions effectively.[45]

PRO TRADE

Since the end of World War II, the western European allies, with the temporary exception of West Germany, have attempted to keep economics and politics separate in the formulation of their trade policy toward the East. They believe that the economic benefits of trade should not be subjugated to extraneous concerns, as long as trade entails no seriously adverse strategic consequences for the West. They have practiced a pro-trade doctrine that the United States has tended to oppose, except when self-interest dictates that it dispose of surplus agricultural commodities.

Beneath the trade for trade's sake view, often, lies a strategic vision. Some pro-traders believe commerce can, in the long run, help stimulate a peaceful and positive evolution of Soviet foreign and domestic policies. Unlike linkage, which would use trade explicitly to mold Soviet behavior, free traders believe that increased economic exchanges, of their own accord, will influence Soviet actions by bringing to the fore more pragmatic leaders and exposing both the Soviet elite and the general populace to Western goods, ideas, and people. Giovanni Agnelli, the Italian automobile magnate who has dealt extensively with the Eastern bloc, notably in the construction of the Togliattigrad automobile plant, sums up this school of thought:

> I believe that on the whole, and within certain limits, trade does indeed encourage the growth, inside the Soviet society, of forces and views naturally oriented toward the pursuit of more peaceful relations with the rest of the world. It is not so difficult to identify, in the Soviet hierarchy of power, those economic and technocratic groups which are in favor of strengthening ties with the West for economic reasons in order to develop Soviet technology and in general to compensate for Soviet economic failures and backwardness. If these people and groups have a certain influence on Soviet policy—and they presumably do have one—theirs is bound to be an influence for detente and peace, rather than for actions leading to a cold war atmosphere.[46]

This "getting to know you" strategy resulted in many contacts between Western businessmen, academics, scientists, and artists, and their Soviet counterparts during the 1970s. Thousands of Soviets worked with Western engineers and technicians on industrial projects in various parts of the country. Human and intellectual interaction of this sort may help to change perspectives favorably. Each side may begin to understand that its adversary is not simply a faceless monster ruling an evil empire, or a greedy imperialist colonizer. Many hard-nosed Western diplomats have testified that personal contact with the Russians made them more open to the idea of seeking a constructive working relationship with the East. Presumably, similar phenomena may occur within the Soviet elite exposed to the West. It is in this manner that the Cold War stereotypes can be gradually erased.

Winston Churchill once remarked that the Kremlin fears the West's friendship no less than its animosity. As Marshall Goldman writes, "The Soviets have learned that the same technology that they bring in to strengthen their military forces may also tend to subvert if not debilitate the moral resoluteness of the population."[47]

Consider, for example, the Kremlin's computer predicament. The scientific and academic communities are begging for microprocessors to help them invent and learn. But the government is concerned that giving large segments of the population access to computers will pose a serious threat to the centralized control of information—a cornerstone of the state's power. In a system where even the most innocuous government statistics are considered secrets, the thought of a computer hack plugging into an official data bank and learning about Moscow's air defense system, the shortfalls of the economic plan, or even about the true infant mortality rate terrifies the authorities. In a country where photocopying machines are kept under watch in the vain hope of preventing their unauthorized use, officials fear that dissidents with high speed printers could easily copy banned literature from computer disks pilfered at home or smuggled in from abroad.[48]

This fear of computerization is a paradigm of the more general dread of change. A number of prominent scholars, notably Richard Pipes, argue that the Soviet elite has no self-interest in reform. The *nomenklatura* is a priviliged class by virtue of its monopoly on power. Since liberalization would require a redistribution, a decentralization, and individualization of power, the elite has nothing to gain, and everything to lose, through reform.

And yet, despite their fear of change, some of the leaders in the Kremlin understand that in order for the Soviet Union to keep up with the rest of the world in the midst of a technological revolution, its citizens must have the freedom to create. Gorbachev, as evidenced by his sermon before the 27th Party Congress and his speeches around the Soviet Union, seems to espouse this view. While it is naive to believe that a well-tailored suit and a facility to charm the world media are the signs of a reformer, Gorbachev's pronouncements indicate his awareness that without a greater degree of freedom there will be no sustained economic progress and the USSR will be condemned to remain forever behind in its competition with the capitalist world.

From the pro-trade perspective, the West has a part to play in helping the Soviet bloc decide between progress and paralysis. By severing commercial ties and attempting to force its allies to do the same, the United States only bolsters those within the Soviet elite who would continue to devote all of their nation's massive resources to a military buildup. By promoting trade and economic interaction, the United States and the West might help bring to the fore those who would rather shift priorities to the civilian economy.

The events of the past decade somewhat dim the luster of the pro-trade theory. Soviet leaders seem all too aware of the potentially subversive effects of trade. Richard Barnet has written that increased trade in the 1970s did not contribute much in the way of internal reform, with the exception of a more lenient, albeit short-term, emigration policy. At the 24th Party Congress on the eve of détente in 1971, Brezhnev declared that his "peace" program would be linked to "an unabating ideological war." The party and the people, he said, "have not tolerated and will not tolerate attempts—no matter what their origin—to blunt our ideological weapons, to stain our banner."[49]

This standard ideological blustering may have been intended for internal consumption while Communist Party doctrine was being compromised every day in an attempt to develop closer economic contacts with the capitalist world. The fact, however, is that despite the Helsinki Accords, which were supposed to guarantee a modicum of human rights, dissidents continued to be harassed, mistreated, and imprisoned, as the tragic experience of Andrei Sakharov testifies. Nor did trade with the West keep the Soviets out of Afghanistan, curb Soviet defense expenditures, prevent Moscow from helping to topple Solidarity, or convince the Soviets not to meddle in Central America, Angola, and other Third World hot spots.

Yet these criticisms of the pro-trade strategy are themselves subject to reservations. First, even limited détente produced positive results, especially within the satellite countries. Under the impact of growing business with the West, Hungary quietly instituted market-oriented economic reforms. Romania implemented a more independent foreign policy. The two Germanys developed stronger and presumably lasting ties. During the 1970s, "far from Russia dominating Western Europe, it was Eastern Europe that began to move closer to the West."[50]

Indeed, the very success of détente in promoting liberal reform within the Eastern bloc countries is reason enough to give that policy another chance. As German scholar Heinrich Vogel points out, the greater the stability between East and West, the greater the margin for maneuver by those living in the satellite countries. Conversely, the greater the tension and suspicion, the smaller the field for autonomous decision making by the satellites.[51] Thus, even if increased trade fails to kindle significant change within the USSR, it should still improve the lot—both materially and politically—of those subjugated to Soviet will.

The short-lived détente also had a positive impact on the Soviet Union. During the thaw and the expansion of economic relations that accompanied it, the Soviets allowed more than half a million Jews and ethnic Germans to leave the USSR. Emigration returned to a trickle when it became clear that the commercial relationship with the United States was at a dead end and that political benefits such as SALT II were not attainable. The fear of compromising all ties with the West may have prevented a Soviet move into Yugoslavia after Marshall Tito's death or an outright invasion of Poland during the Solidarity uprisings.[52] Such hypotheses are impossible to substantiate, a fact that points to a central drawback to pro-trade: the difficulty of proving its successes.

More fundamentally, pro-traders have trouble proving their own case because a policy of economic détente was never fully implemented. Commercial contacts between East and West, held in check by the U.S. Congress, did not rise to the considerable levels necessary to inculcate restraint in Soviet behavior. Moscow's adventurism in the late 1970s came at a time when the promise of meaningful economic interaction had gone sour. In sum, the trade gambit and détente itself were not given the opportunity to work because they were pursued by both sides with "one foot on the accelerator and the other on the brake."[53]

Despite its checkered beginnings and the current disappointment over détente, the Siberian pipeline may yet prove to be a good example of the pro-trade strategy in action. Both East and West invested heavily in a project that promises each side potentially important long-term benefits. A thick strand in the web of commerce, the pipeline should ensure that engineers, businessmen, economic planners, and government leaders maintain a dialogue as they seek to continue the supply of energy, on the one hand, and the inflow of hard currency and technology, on the other. By promoting a healthy degree of economic interdependence between the antagonistic systems, projects such as the Siberian pipeline pull the East-West relationship one step back from catastrophic conflict and push it one step ahead toward cautious coexistence and even cooperation.

10

A Crisis in Context

It sometimes appears that only bad news is news within the Atlantic alliance. Since the pipeline crisis, a series of other disputes have upset the Western partners. Tens of thousands of Europeans protested the deployment of Pershing II nuclear missiles, as television cameras captured the ironic image of demonstrators picketing outside of NATO bases created to protect them. The United States, citing unfair dumping practices, imposed stiff tariffs on European steel exports and subsidized the worldwide sales of its agricultural products in the face of allegedly unfair European competition. When Washington decided to retaliate against Libya for its terrorist activities, a number of allies, notably France, refused even the minimal cooperation of allowing American bombers to overfly their territory. In this context, the bitter allied acrimony over the pipeline seems like just one of many temporary irritants that have periodically plagued the alliance during its 40-year existence.

A number of other developments lend more specific support to the argument that the pipeline dispute was overblown. Both the hopes and fears concerning the project proved to be exaggerated: it neither guaranteed Europe's energy future nor posed a serious threat to Western security. The invisible hand of the market so dear to President Reagan has, in time, accomplished far more than his uncharacteristic and heavy-handed intrusion into the business life of friendly countries and the arena of East-West trade. The current abundance of oil and Norway's decision to open new and vast gas fields make Soviet gas much less important a factor to the EEC than it appeared in 1980 or even 1982. Dependence on the Soviet Union for energy, never in fact a serious problem, has become virtually a moot issue. Similarly, Soviet earnings from the pipeline will be far below the $10 to $15 billion a year that was once projected. The $2 billion or so that Moscow does receive over the next few years will be poured back into energy production. There will be little excess to reverse the deterioration of the Soviets' hard currency reserves.

Against this background, it might be tempting to view the alliance crisis over the pipeline as an unpleasant but insignificant melodrama that should now be relegated to a footnote in history books. That would be a mistake. What sets the pipeline crisis apart from other alliance squabbles is the fact that it manifested a fundamental difference between the United States and the western Europeans with regard to their policies for East-West relations. These differences date back to the creation of the Western alliance, but came to the fore only in 1982. While the pipeline dispute itself was settled, the larger West-West antagonism of which the crisis was but a symptom remains.

America's allies in Europe have built their economies to a large extent around foreign trade. The Soviet bloc economies, and notably that of the USSR, are especially complementary to that of western Europe. A classic equation of mutual interest resulted: the East exports the natural resources that the western Europeans lack and in exchange imports from them the finished goods and high technology that it needs and is unable to produce efficiently.

Foreign trade plays a much less significant role in the American economy. Moreover, complementarity between the United States and the Soviet bloc is low—the United States has little need for exports from the East. Despite the existence within the Soviet Union and eastern Europe of potentially huge markets for American products—agricultural and industrial—and for American technology, trade policy toward the Soviet bloc has been subjugated to geopolitical and military considerations.

After World War II, the natural propensity for trade between western Europe and the East was stifled by the Cold War and the United States' dominant position within the alliance. Following its economic reconstruction aided by the Marshall Plan, Europe acquired a more assured voice in the formulation of trade policy toward the Soviet bloc. By then, the United States had also begun to develop an interest in the benefits of commerce with the Soviet Union. During the 1970s, détente produced a dramatic increase in trade between East and West. But as Washington's attitude toward détente and trade soured at the end of the decade, the Europeans continued to reap the rewards of commercial interaction. The growing allied divergence over East-West trade policy, more pronounced than at any time since World War II, set the backdrop for the pipeline crisis.

In 1986 the West remains deeply divided over the trade issue. The Europeans continue to believe in the economic and political benefits of East-West commerce; Washington still views such ties with great suspicion. Attempts to paper over the cleavage with studies and diplomatic devices—what the French call ''drowning the fish''—have not made the problem go away. In fact, the intra-alliance antagonism is almost certain to produce more disputes like the one surrounding the pipeline, as Europe's export markets continue to shrink and new opportunities for trade with the Eastern bloc open.

Moscow has clearly expressed its desire to expand commerce with the West, particularly the United States. Gorbachev made this plain at the 27th Party Congress and to the numerous American businessmen and government leaders with whom he has met. High-level Soviet representatives have openly referred to the possibility of increased Jewish emigration in return for a relaxation of U.S. trade restrictions. Foreign Ministry officials have spoken of their country's interest in joining the International Monetary Fund, GATT, and other established institutions of world trade and finance.

The Kremlin's ambitious plans for the reform and expansion of the ailing Soviet economy, including the modernization of an archaic industrial base, depend at least in part upon increased trade. To finance that trade, Moscow must first ensure itself an adequate supply of hard currency. Yet falling oil production and prices have reduced Soviet hard currency earnings by 30 percent. Plans to sell Finland and Yugoslavia nuclear reactors—and thus earn precious hard currency— were put on hold following the Chernobyl accident. Energy sources otherwise destined for the West may have to be used to compensate for a loss of nuclear power generation. In the near term, Moscow can only bring in more currency by exporting its ''excess'' resources, such as natural gas, to the West.

Meanwhile, Soviet requirements for agricultural and industrial goods and technology from the capitalist countries continue to grow. If the production of natural gas, hydroelectricity, and coal is to be expanded in an efficient manner, Western technology will have to play a significant role. The Chernobyl disaster affected some of the most fertile land in the Ukraine—grain and livestock imports may have to increase accordingly. All of these facts indicate that the Soviets will be eager to do business with the West.

It takes two to trade. The Europeans are willing, the United States opposed, or hesitant at best. For this reason, a dispute within the West is almost certain to arise once again, perhaps over the very next megaproject. But whether it will be contained, as the pipeline crisis was, and at what price, is anyone's guess. This danger facing the alliance can be averted only if the lessons of the pipeline chapter are understood—and acted upon.

The United States and western Europe are partners in a competitive relationship among themselves and in an adversary relationship with the Communist world that will continue into the indefinite future. From time to time, unilateral policies, whether self-serving or sincerely meant to benefit the alliance as a whole, only manage to harm Western economies and breed disunity within NATO. Too often, the Soviet Union and its satellites stand idly on the sidelines, watching the allies attempt to hang each other with a rope of their own making.

This problem results from the very nature of the alliance and is, to some extent, inevitable. A grouping of independent democracies united in a common strategic cause for almost four decades is an unusual and precious phenomenon, but one with an Achilles' heel. In peacetime, differences emerge when a nation pursues its own particular vision of the alliance interest in conflict with the perception of its partners or when it ignores the group interest to serve its own. The very characteristic of the group relationship we most cherish—the sovereignty of each member—tends to emerge as a constant weakness in our competition with the Communist world. This produces frustration, particularly in the United States with its superpower responsibilities. Yet a fact of life in the Western alliance is that no one member can issue marching orders to the others. To do so would be to make the alliance indistinguishable from the Warsaw Pact, which is based on a relationship of intimidation and domination decreed by the Soviet doctrine of limited sovereignty. Alliance policy, in short, must be the product of compromise, not coercion.

Looking back at the pipeline crisis, one can distinguish the broad outlines for a workable compromise on East-West trade policy. First, the United States must accept one given: western Europe will not forego the benefits of trade with the East in conditions short of war or acute tension. "No matter what their orientation—conservative, centrist or leftist—leaders across the Atlantic see East-West trade with the same eyes," says Helmut Schmidt. "Maintaining economic relations with the East is an essential part of European policy. We have been trading with the Russians for hundreds of years; we are all part of an inter-European trade network. The bottom line is simple: We will not let the United States dictate this aspect of our economic policy."[1]

The European view that Schmidt expresses so forcefully cannot be attributed merely to self-interest. It also reflects the understanding of America's partners that the NATO alliance they joined was meant to be a shield not a sword. NATO's charter describes a defensive alliance. The idea of using the organization as a weapon to strangle the USSR militarily and economically—if that were possible—runs counter to its founders' intent. If the conception of NATO as a structure designed to assure collective security is still valid, it follows that aggressive policies of trade denial are not acceptable.

Whatever the basis for European attitudes, one fact is clear: for the West to avoid another pipeline crisis, the United States must at least tolerate, and perhaps even move in the direction of, western Europe's pro-trade approach. Yet continued European insistence on trading with the East is not in itself a sufficient reason for the United States to join the pro-trade consensus. What should be determinative is the fact that alternative strategies have simply not worked in the past and hold no promise of future success. Return to a policy that accepts economic ties with the Soviet bloc is a pragmatic compromise compatible with the West's security requirements that the United States should move to embrace.

Trade denial has failed and will continue to do so. In a world of interdependent economies in which the United States is far from being the only supplier of industrial equipment and technology, Washington alone is incapable of denying the Soviet Union sophisticated products and know-how. Sweeping attempts to curb the eastward flow of anything but the most strategic goods serve mostly to constrain West-West trade and to hinder the development of the very same technology that is the source of American economic and military strength. By the same token, such attempts are harmful to the United States' business potential in the world market. They lead to losses in foregone eastbound trade, which the Europeans and Japanese know how to turn into profit. Further, they discourage Western industries from buying American licenses, patents, and products lest Washington proscribe their own exports. Even if we could keep more of our technology out of the Soviets' hands and get our allies to do the same, it is wrong to believe that the USSR could not, in the end, develop indigenously what it most needs. This is the lesson of all long-lasting boycotts and embargoes.

Heavy-handed linkage between the privilege of American trade and good Soviet behavior performs little better. Soviet dependence on Western goods and markets is too low to achieve the desired results; also, sanctions impose costs on the wielder of the trade club often greater than those incurred by the target. At the same time, America's allies object to Washington's habit of using embargoes to achieve broad foreign policy goals, rather than for clear-cut reasons of national security. Without allied cooperation, linkage is likely to be doomed in advance. Moreover, swinging the club wide and far often produces embarrassment, not results. Washington tends to connect sanctions to specific changes it seeks in the behavior of the Soviet Union or other countries. These changes rarely follow. If we stop selling grain to Moscow because Soviet troops are in Kabul, or if we embargo the sale of pipeline technology to the USSR because martial law was imposed in Poland, how can we resume those sales and lift the embargoes when the troops and martial law remain?[2] In the end, Washington appears weak and ineffective, as the pipeline crisis painfully demonstrated.

It may be wishful thinking to believe that expanded economic relations will produce positive change in the Kremlin's foreign and domestic policies. But this approach deserves a fairer and longer trial than it received during the 1970s. At worst, the West will benefit economically from its interaction with the East while the Soviet satellites should enjoy a wider margin for maneuver. Paradoxically, by moving in the direction of its allies on trade policy, Washington should be able to achieve worthwhile strategic objectives that have so far been frustrated. Export controls limited to items with obvious military applications would be easier to enforce than an open-ended embargo covering products and know-how with farfetched dual usages. In a more liberal trading climate, allied cooperation in denying the Soviets of truly critical technology with direct military significance undoubtably would increase. The Europeans would be more easily persuaded to curtail heavily subsidized credits for East bloc countries. Severe allied friction, like that resulting from the pipeline crisis, would be dispelled.

From the perspective of American strategic interests, much can be said for a policy that engenders stability and comity within the alliance. George Kennan has written that:

> The problem of containment is basically a problem of the reactions of people within the non-communist world. It is true that this condition depends upon the maintenance by ourselves and our allies, at all times, of an adequate defense posture, designed to guard against misunderstandings and to give confidence and encouragement to the weak and the fainthearted. But so long as that posture is maintained, the things that need the most to be done to prevent the further expansion of Soviet power are not, so far as we are concerned, things we can do directly in our relations with the Soviet government; they are things we must do in our relations with the peoples of the non-communist world.[3]

By promoting a more harmonious alliance, rather the one divided over an issue as fundamental as East-West trade relations, the West will be in a better position to meet the challenges posed by its adversaries.

Of course, there is little doubt that the West has the will and the means to counter direct aggression. The overriding mutual interest in the survival of our value system supercedes all temporary blindness to group interest. Yet the issues facing the West are usually more subtle than direct aggression. Whether the problem is deploying new nuclear weapons in western Europe, maintaining fair trading practices, cooperating in the fight against terrorism, or establishing a commercial relationship with the Eastern bloc that meets strategic concerns, a minimal amount of goodwill and solidarity is necessary to overcome incompatible short-term interests. Repeated disputes like the one over the Siberian pipeline will make that goodwill and solidarity, both in times of a crisis with an outside force and in the day-to-day life of the alliance, that much more difficult to attain.

Just as it is wise to seek to understand the motivations and interests of an adversary, so it is imperative to try to comprehend the viewpoints and constraints of one's partners. If the Siberian pipeline crisis teaches us anything, it is that the Western alliance must look inward, and not simply outward, if it is to remain secure.

Notes

CHAPTER 1

1. George Kennan, "Current Problems of Soviet-American Relations," Naval Academy, May 9, 1947, *Kennan Papers,* Box 17 as quoted in Barton Gellman, *Contending with Kennan: Toward a Philosophy of American Power* (New York: Praeger, 1984), p. 138.

CHAPTER 2

1. Wilhelm Christians, interview, Düsseldorf, West Germany, June 1985.
2. International Energy Agency, *Energy Policies and Programmes of IEA Countries—1982 Review* (Paris: OECD/IEA, 1983), p. 401; Angela E. Stent, *Soviet Energy and Western Europe* (New York: Praeger, 1982), p. 104.
3. Robert J. Lieber, *The Oil Decade: Conflict and Cooperation in the West* (New York: Praeger, 1983), p. 17.
4. Ibid., p. 16.
5. *New York Times,* February 14, 1982, p. E5.
6. See John L. Gaddis, *Strategies of Containment* (New York: Oxford University Press, 1982), pp. 178–79; Walter LaFeber, *America, Russia and the Cold War* (New York: John Wiley and Sons, 1980), pp. 192–97; Dankert A. Rustow, *Oil and Turmoil* (New York: Norton, 1982), pp. 83–84; Steven L. Spiegel, *The Other Arab-Israeli Conflict* (Chicago: University of Chicago Press, 1985), pp. 71–82.
7. Rustow, *Oil and Turmoil,* pp. 249–51; Walter J. Levy, "Oil and the Decline of the West," *Foreign Affairs,* Summer 1980, pp. 1009–11; Robert W. Tucker, "American Power and the Persian Gulf," *Commentary,* November 1980; and M. H. Blinken, letters to *New York Times,* June 21, 1978, December 5, 1978, and December 4, 1979.
8. Helmut Schmidt, interview, New York City, October 1985.
9. Hanns W. Maull, *Natural Gas and Economic Security: Problems for the West* (Paris: Atlantic Institute, 1981), p. 9.
10. "The European Community's Energy Strategy," *European File* no. 8 (1982):1.
11. Organization for Economic Cooperation and Development, *Energy Prospects to 1985* (Paris: OECD, 1974), introduction; Maull, *Natural Gas,* pp. 8–13.
12. Jonathan B. Stein, *The Soviet Bloc, Energy and Western Security* (Lexington, MA: Lexington Books, 1983), p. 176.
13. Maull, *Natural Gas,* p. 15.
14. Ibid., p. 18.
15. Stein, *The Soviet Bloc,* p. 77.
16. Royal Institute of International Affairs, *Policies and Strategies of Those Countries Which Export Natural Gas to Western Europe* (London: RIIA, 1983), p. 3.
17. Maull, *Natural Gas,* p. 26.
18. Ibid., p. 16 and p. 20, table 2.

19. *Petroleum Intelligence Weekly,* June 18, 1984.

20. International Energy Agency, *Natural Gas: Prospects to 2000* (Paris: OECD/IEA, 1982), p. 95.

21. Ibid.

22. Hanns W. Maull, *Europe and World Energy* (London: Butterworth, 1980), ch. 2; Organization for Economic Cooperation and Development, *Economic Surveys: Norway* (Paris: OECD, 1980), pt. III; IEA, . . . *Prospects to 2000,* p. 93; *New York Times,* June 3, 1986, p. 1.

23. *New York Times,* July 8, 1985, p. D1.

24. IEA, . . . *Prospects to 2000,* p. 93; *New York Times,* June 3, 1986, p. 1; *Wall Street Journal,* June 4, 1986, p. 35; *Libération,* June 4, 1986, p. 11.

25. Maull, *Natural Gas,* p. 20.

26. IEA, . . . *Prospects to 2000,* p. 112.

27. Ibid., p. 110.

28. Maull, *Natural Gas,* p. 20.

29. Sonatrach/Bechtel Corporation, *The Hydrocarbon Development Plan of Algeria: Financial Projections 1975-2000* (San Francisco: Bechtel Corporation, 1978), introduction.

30. IEA, . . . *Prospects to 2000,* p. 12.

31. Michael R. Meyer, "Tapping the Sahara's Riches," *Newsweek,* July 18, 1983, p. 38.

32. I. F. Elliott, *The Soviet Energy Balance* (New York: Doubleday, 1974), pp. 12-18.

33. Jonathan P. Stern, "Western Forecasts of Soviet and Eastern European Energy over the Next Two Decades (1980-2000)," in U.S. Congress, Joint Economic Committee, *Energy in Soviet Policy* (Washington, D.C.: U.S. Govt. Printing Office, 1979-80), p. 32.

34. Jonathan P. Stern, *Soviet Natural Gas Development to 1990* (Lexington, MA: Lexington Books, 1980), p. 28, Table 2-5.

35. A. B. Smith, "Soviet Dependence on Siberian Resource Development," in U.S. Congress, *Soviet Economy in a New Perspective* (Washington, D.C.: U.S. Govt. Printing Office, 1980), p. 490, Table 4; *Guidelines for the Economic and Social Development of the USSR for 1986-1990 and the Period Ending in 2000* (Moscow: Novosti Press Agency, 1985), p. 47.

36. Stern, *Soviet Natural Gas,* p. 35.

37. Ibid., p. 36.

38. Frank Holzman, *International Trade Under Communism—Politics and Economics* (New York: Basic Books, 1976), p. 163.

39. Angela E. Stent, *From Embargo to Ostpolitik* (Cambridge: Cambridge University Press, 1981), p. 165. The use of VOEST as a go-between reflected Soviet ambivalance about dealing with the Germans in the wake of the 1962 pipe embargo. But Moscow was very satisfied with the deal, and its reluctance to work directly with the Germans vanished.

40. Ibid., p. 164.

41. Ibid., pp. 166-67; International Monetary Fund, Bureau of Statistics, *International Financial Yearbook, 1983* (Washington, D.C.: IMF, 1984), p. 67.

42. Stent, *Embargo to Ostpolitik,* pp. 166-67.

43. International Monetary Fund, *Yearbook,* p. 67.

44. Richard Barnet, *The Giants: America and Russia* (New York: Simon and Schuster, 1977), p. 147.

45. Ibid.

46. Stern, *Soviet Natural Gas,* pp. 109–11.

47. Barnet, *The Giants,* p. 147.

48. Peter G. Peterson, "U.S.-Soviet Commercial Relations in a New Era," *U.S. Department of Commerce Report,* August 1972, p. 5.

49. *Oil and Gas Journal,* May 10, 1976, p. 120.

50. Giovanni Agnelli, "East-West Trade: A European View," *Foreign Affairs,* Summer 1980, p. 1023. There was some speculation that the United States was intent on stopping the deal for the benefit of the Alaskan fields.

51. Stern, *Soviet Natural Gas,* p. 79.

52. "Plus de Pétrole Iranien pour la France," *Le Monde,* September 15, 1980; IEA, . . . *Prospects to 2000,* p. 117.

53. Stern, *Soviet Natural Gas,* p. 105, Table 5-4.

54. Thane Gustafson, *Soviet Negotiating Strategy: The East-West Gas Pipeline Deal, 1980–1984* (Santa Monica, CA: Rand, 1985), p. 10.

55. Ed A. Hewett, *Energy, Economics and Foreign Policy in the Soviet Union* (Washington, D.C.: Brookings Institution, 1984), p. 76.

56. H. Erich Heinemann, "Overflow Swamps OPEC's Oil Prices," *New York Times,* February 27, 1983.

57. *New York Times,* February 14, 1982, p. E5.

58. Stein, *The Soviet Bloc,* p. 59.

59. *New York Times,* September 29, 1981.

60. John Tagliabue, "Bonn Needs the Business Even More Than the Gas," *New York Times,* August 16, 1981.

61. *Financial Times,* November 16 and December 31, 1982.

62. Jonathan P. Stern, "Specters and Pipe Dreams," *Foreign Policy,* Fall 1982, p. 30.

63. Stein, *The Soviet Bloc,* pp. 72–73; Heinrich Vogel, "Political-Security Implications of East-West Trade" to be published in *Challenges to the North Atlantic Alliance.*

CHAPTER 3

1. *Financial Times,* October 15, 17; December 1, 11, 16, 18, 1980; Thane Gustafson, *Soviet Negotiating Strategy: The East-West Gas Pipeline Deal, 1980–1984* (Santa Monica, CA: Rand, 1985), p. 17.

2. Gustafson, *Negotiating Strategy,* p. 17.

3. Gordon Crovitz, *Europe's Siberian Gas Pipeline: Economic Lessons and Strategic Implications* (London: Institute for European Defence and Strategic Studies, 1983), p. 12; International Monetary Fund, Bureau of Statistics, *International Financial Yearbook, 1983* (Washington, D.C.: IMF, 1984). p. 67.

4. Wilhelm Christians, interview, Düsseldorf, West Germany, June 1985.

5. Crovitz, *Siberian . . . Pipeline,* p. 12; IMF, *Yearbook,* p. 67.

6. Gustafson, *Negotiating Strategy,* p. 20.

7. Ed A. Hewett, *Energy, Economics and Foreign Policy in the Soviet Union* (Washington, D.C.: Brookings Institution, 1984), p. 199.

8. Crovitz, *Siberian . . . Pipeline*, p. 12.

9. Axel Lebahn, "The Yemal Gas Pipeline from the USSR to Western Europe in the East-West Conflict," *Aussen Politik* 34 (3rd quarter 1983).

10. Christians, interview.

11. Jean-François Revel, *Comment les Démocraties Finissent* (Paris: Grasset, 1983), p. 83.

12. Crovitz, *Siberian . . . Pipeline*, p. 33; IMF, *Yearbook*, p. 67; Giovanni Agnelli, "East-West Trade: A European View," *Foreign Affairs*, Summer 1980, p. 1023.

13. Gustafson, *Negotiating Strategy*, p. 19.

14. Christians, interview.

15. Gustafson, *Negotiating Strategy*, p. 23.

16. Jonathan B. Stein, *The Soviet Bloc, Energy and Western Security* (Lexington, MA: Lexington Books, 1983), p. 63; *New York Times*, September 29, 1981; *Financial Times*, October 30, 1981.

17. Stein, *Soviet Bloc*, p. 63; *Financial Times*, October 30, 1981.

18. Lionel Olmer, quoted in Office of Technology Assessment, *Technology and East-West Trade: An Update* (Washington, D.C.: U.S. Govt. Printing Office, 1983), p. 57.

19. Samuel Pisar, *Coexistence and Commerce: Guidelines for Transactions Between East and West* (New York: McGraw-Hill, 1970), pp. 312–13.

20. Ibid., pp. 273–79.

21. "Paying the Piper," *Wall Street Journal*, September 30, 1983, p. 30; *Financial Times*, December 12, 1980.

22. Stein, *Soviet Bloc*, pp. 60–61.

23. B. A. Rahmer, "Big Gas Deal with West Europe," *Petroleum Economist*, January 1982, p. 13; Crovitz, *Siberian . . . Pipeline*, p. 24; Gustafson, *Negotiating Strategy*, p. 31; *Financial Times*, October 21, 1981.

24. *Financial Times*, December 10, 1981.

25. Gustafson, *Negotiating Strategy*, p. 32.

26. Rahmer, "Big Gas Deal," p. 13.

27. Crovitz, *Siberian . . . Pipeline*, p. 24.

28. *Petroleum Intelligence Weekly*, March 15, 1982, p. 8; Gustafson, *Negotiating Strategy*, pp. 31–34.

29. *Le Monde*, December 8, 1984.

30. *Le Monde*, June 7 and 27, 1985; *Libération* (Paris), June 7 and 27, 1985; *Financial Times*, September 4, 1984.

31. International Energy Agency, *Natural Gas: Prospects to 2000* (Paris: OECD/IEA, 1982), pp. 38–39; *Le Monde*, June 30, 1985, p. 14.

32. William Safire, "We Told Them So," *New York Times*, March 27, 1983.

33. Christians, interview.

34. Gustafson, *Negotiating Strategy*, p. 31.

35. Werber Nowak and Kurt Rippholz, "Surveying a Pipeline from Each End," *Wall Street Journal*, January 12, 1984, p. 25.

36. "Gaz Soviétique: De Surcroît en Surplus, C'est la Surchauffe," *Libération* (Paris), August 5, 1983.

37. International Energy Agency, *Energy Policies and Programmes of IEA Countries—1982 Review* (Paris: OECD/IEA, 1983), introduction.

38. Alan Manne and William Nordhaus, "Using Soviet Gas to Keep OPEC Reeling," *New York Times,* September 22, 1985, p. F3.

39. IEA, . . . *Prospects to 2000*, p. 32.

40. International Energy Agency, *OECD Quarterly Oil Statistics—1st Quarter, 1983* (Paris: OECD/IEA, 1983).

41. Charles D. Masters, "World Petroleum Resources—A Perspective," Open File Report 85-248 (U.S. Geological Survey, 1985), p. 13.

42. Marshall Goldman, interview, Cambridge, Massachusetts, fall 1983.

CHAPTER 4

1. Interview, official in the Department of Defense who wished to remain anonymous.

2. Richard Pipes, interview, Cambridge, Massachusetts, fall 1983.

3. Jonathan P. Stern, *Soviet Natural Gas Development to 1990* (Lexington, MA: Lexington Books, 1980), p. 141.

4. International Energy Agency, *Natural Gas: Prospects to 2000* (Paris: OECD/IEA, 1982), p. 41.

5. Stern, *Soviet Natural Gas,* p. 141.

6. Boyce I. Greer, "Soviet Natural Gas Exports and Western Security: The Yamburg Natural Gas Pipeline" (unpublished paper, Harvard University, 1981), p. 7.

7. Hanns W. Maull, *Natural Gas and Economic Security* (Paris: Atlantic Institute, 1981), p. 28.

8. Jonathan B. Stein, *The Soviet Bloc, Energy and Western Security* (Lexington, MA: Lexington Books, 1983), p. 78.

9. Stern, *Soviet Natural Gas,* p. 141.

10. Arthur J. Klinghoffer, *The Soviet Union and International Oil Politics* (New York: Basic Books, 1979), pp. 144-55.

11. *Petroleum Intelligence Weekly,* April 16, 1979; Stern, *Soviet Natural Gas,* p. 139.

12. "Scargill's Big Brother" (editorial), *Wall Street Journal,* November 20, 1984.

13. Maj. Gen. Richard X. Larkin and Edward M. Collins, Defense Intelligence Agency, statement before the Joint Economic Committee, Subcommittee on International Trade, Finance, and Security Economics, in *Allocation of Resources in the Soviet Union and China—1981* (Washington, D.C.: U.S. Govt. Printing Office, 1981), p. 48; Pipes, interview.

14. Scenario related by the author in "Pipeline to Prosperity?" *Harvard Crimson,* February 12, 1982, p. 2, from an article in *L'Express* (Paris).

15. Stern, *Soviet Natural Gas,* p. 140.

16. Helmut Schmidt, interview, New York City, October 1985.

17. Bill Paul, "Gas Glut May Trouble Soviet Pipeline," *Wall Street Journal,* January 19, 1984.

18. Larkin/Collins statement, p. 48.

19. *Wall Street Journal,* March 16, 1985.

20. *Wall Street Journal,* September 28, 1984, p. 31.

21. Ed A. Hewett, *Energy, Economics and Foreign Policy in the Soviet Union* (Washington, D.C.: Brookings Institution, 1984), p. 198.

22. Central Intelligence Agency, *Prospects for Soviet Oil Production: A Supplemental Analysis* (Washington, D.C.: CIA, 1977), p. 32; Marshall Goldman, *The Enigma of Soviet Petroleum: Half Empty or Half Full?* (London: Allen and Unwin, 1980), p. 109.

23. Theodore Shabad, "Soviet Production of Oil Shows an Unusual Drop," *New York Times,* March 4, 1984, p. 14; Larkin/Collins statement, p. 48.

24. "Grain Pact Signed: U.S. Assures Soviet of Steady Supply," *New York Times,* August 26, 1983, p. 1.

25. For two excellent descriptions of Soviet economic problems, see Marshall Goldman, *USSR in Crisis: The Failure of an Economic System* (New York: W. W. Norton, 1983); and Samuel Pisar, *La Ressource Humaine* (Paris: Lattes, 1983).

26. Maull, *Natural Gas,* p. 47.

27. Leslie Dienes and Theodore Shabad, *The Soviet Energy System: Resources Use and Policies* (New York: John Wiley and Sons, 1979), p. 67.

28. Beverly Crawford and Stefanie Lenway, "Decision Modes and International Regime Change: Western Collaboration on East-West Trade and the Trans-Siberian Pipeline Dispute" (paper delivered to the American Political Science Association, Chicago, 1983), p. 30.

29. U.S. Congress, Joint Economic Committee, *Soviet Pipeline Sanctions: The European Perspective* (Washington, D.C.: U.S. Govt. Printing Office, 1982), p. 22.

30. Thane Gustafson, *Soviet Negotiating Strategy: The East-West Gas Pipeline Deal, 1980–1984* (Santa Monica, CA: Rand, 1985), p. 12.

31. Pisar, *Ressource Humaine,* pp. 171–72.

32. Giovanni Agnelli, "East-West Trade: A European View," *Foreign Affairs,* Summer 1980, p. 1023.

33. *New York Times,* February 8, 1982.

34. Pipes, interview.

35. Central Intelligence Agency, "Outlook for the Siberia-to-Western Europe Natural Gas Pipeline," Sov 82-10120, Eur 82-10078, August 1982.

36. David Buchan, "Russia May Use Own Gas Turbines on Siberian Gas Pipeline," *Financial Times,* July 30, 1982, p. 2; Office of Technology Assessment, *Technology and East-West Trade: An Update* (Washington, D.C.: U.S. Govt. Printing Office, 1983), p. 68; Hewett, *Energy . . . Soviet Union,* p. 79.

37. Fabio Basagni, interview, Atlantic Institute, Paris, summer 1983.

38. See Avraham Shifrin, *The First Guidebook to Prisons and Concentration Camps of the Soviet Union* (New York: Bantam, 1982); also Gordon Crovitz, *Europe's Siberian Gas Pipeline* (London: Institute for European Defence and Strategic Studies, 1983), p. 39.

39. U.S. Department of State, *Report to the U.S. Congress on Forced Labor in the USSR* (Washington, D.C.: U.S. Govt. Printing Office, 1983), pp. 1–4.

40. Radio Liberty Research, *Soviet Political Prisoners Today: Types of Punishment and Places of Detention* (Washington, D.C.: Radio Liberty Research, 1984); Amnesty International, *Prisoners of Conscience in the USSR: Their Treatment and Condition* (London: Amnesty International, 1981), introduction.

41. Ibid.

42. "Forced Labour in the Soviet Union," *Background Briefs* (London), February 1983, pp. 1–7.

43. Yuri Below and Tamara Waldeyar, *Forced Labor on the Siberian Gas Pipeline* (Frankfurt: International Society for Human Rights, 1982); *Washington Post,* September 22, 1982, p. A14.

44. Crovitz, *Siberian Gas Pipeline,* p. 39.

45. "Prisonniers Sibériens: Mitterrand Attend Toujours Son Rapport," *Libération* (Paris), December 2, 1983, p. 8.

CHAPTER 5

1. George F. Kennan, *Russia and the West Under Lenin and Stalin* (Boston: Little, Brown, 1961), p. 187; Giovanni Agnelli, "East-West Trade: A European View," *Foreign Affairs,* Summer 1980, p. 1018.

2. Quoted in Richard H. Ullman, *The Anglo-Soviet Accord: Anglo-Soviet Relations, 1917–1921* (Princeton: Princeton University Press, 1973), p. 37.

3. Ibid.

4. Marshall Goldman, "Autarky or Integration: The USSR and the World Economy," in U.S. Congress, Joint Economic Committee, *Soviet Economy in a New Perspective* (Washington, D.C.: U.S. Govt. Printing Office, 1976), pp. 82ff.; Samuel Pisar, *Coexistence and Commerce: Guidelines for Transactions Between East and West* (New York: McGraw-Hill, 1970), p. 36.

5. Alec Nove, *East-West Trade: Problems, Prospects, Issues* (Washington, D.C.: Praeger, 1978), p. 89; Richard J. Barnet, *The Giants: America and Russia* (New York: Simon and Schuster, 1977), p. 137.

6. Pisar, *Coexistence and Commerce,* pp. 75–76.

7. Walter LaFeber, *America, Russia and the Cold War* (New York: John Wiley and Sons, 1980), pp. 10–12. For an overview of revisionist versus orthodox explanations for the Cold War, see Adam Ulam, *The Rivals* (New York: Oxford University Press, 1971), pp. 82–83, 93–97.

8. For an account of postwar international events in the area of East-West relations, see LaFeber, . . . *Cold War;* A. W. DePorte, *Europe Between the Superpowers* (New Haven: Yale University Press, 1979); and John L. Gaddis, *Strategies of Containment* (New York: Oxford University Press, 1982).

9. Adam Ulam, *Dangerous Relations: The Soviet Union in World Politics, 1970–1982* (New York: Oxford University Press, 1983), p. 13; DePorte, *Europe . . . Superpowers,* pp. 163–64.

10. U.S. Congress, House, Committee on Banking and Currency, *Hearings on H.R. 4293 to Extend the Export Control Act of 1949* (Washington, D.C.: U.S. Govt. Printing Office, 1969), p. 4.

11. Frank Holzman, *International Trade Under Communism—Politics and Economics* (New York: Basic Books, 1976), p. 67.

12. Fabio Fabbri, "The Balance of Economics and Politics in East-West Trade: The United States and Western Europe in East-West Economic Relation" (unpublished paper, Harvard University, Center for International Affairs, 1983), p. 5.

13. Gaddis, *Strategies,* pp. 37–85; LaFeber, . . . *Cold War,* pp. 59–74; and Ulam, *Dangerous Relations,* pp. 14–16. When the plan was first proposed, the United States

offered assistance to the Soviet Union, Poland, and Czechoslovakia. The Soviets refused aid in the name of all three.

14. Pisar, *Coexistence and Commerce,* p. 132.

15. Office of Technology Assessment, *Technology and East-West Trade* (Washington, D.C.: U.S. Govt. Printing Office, 1979), pp. 113-15.

16. Pisar, *Coexistence and Commerce,* p. 131.

17. Marie LaVigne, *Les Relations Economiques Est-Ouest* (Paris: Presses Universitaires de France, 1979), p. 67.

18. Fabbri, "Balance of Economics," p. 7; Pisar, *Coexistence and Commerce,* pp. 98-106.

19. Holzman, *International Trade,* p. 136. European measures included tariff penalties, quantitative restrictions, and outright prohibition.

20. Richard Rosencrance, *The Rise of the Trading State* (New York: Basic Books, 1985); Pisar, *Coexistence and Commerce,* pp. 46-47.

21. Angela Stent, *From Embargo to Ostpolitik: The Political Economy of West German-Soviet Relations, 1955-1980* (Cambridge: Cambridge University Press, 1981), pp. 239ff. Stent notes several attempts by West Germany to use linkage. *Ostpolitik* means, literally, "East policy," and refers to West Germany's policies for dealing with the Soviet bloc. The end goal of *Ostpolitik* is the reunification of Germany.

22. The boycott problem, difficult to imagine today, should not be underestimated. For an account of several boycotts related to trading with Communist nations and their effectiveness, see Pisar, *Coexistence and Commerce,* pp. 79-87.

23. Jonathan Rosenblatt, *East-West Trade in Technology: A Purpose in Search of a Policy* (Washington, D.C.: Institute for International Economics, 1980), p. 43.

24. Stent, *Embargo to Ostpolitik,* pp. 79-83.

25. Philip Hanson, *Trade and Technology in Soviet-Western Relations* (New York: Columbia University Press, 1981), pp. 92-93.

26. DePorte, *Europe . . . Superpowers,* p. 176.

27. For more detail on "mutual assured destruction," see Gaddis, *Strategies,* pp. 219-20. See also Raymond Aron, *La République Impériale* (Paris: Fayard, 1973), pp. 80ff., for a discussion of the effect of the United States' change in strategic thinking on the Europeans.

28. Wolfgang F. Hanrieder and Graeme P. Auton, *The Foreign Policies of West Germany, France and Britain* (Englewood Cliffs, NJ: Prentice-Hall, 1980), pp. 128-30.

29. Samuel Pisar, *La Ressource Humaine* (Paris: Lattes, 1983), p. 243.

30. Fabbri, "Balance of Economics," pp. 24-25.

31. Stent, *Embargo to Ostpolitik,* pp. 152-53; Hanrieder and Auton, *Foreign Policies,* pp. 60-61. The latter points to domestic constraints on the British government in the form of conservative public opinion that prevented it from pursuing a more adventuresome policy toward the East.

32. Eugene Zeleski and Helgegard Wienert, *Technology Transfer Between East and West* (Paris: OECD, 1980), p. 46.

33. Gunnar Adler-Karlsson, *Western Economic Warfare 1947-1967: A Case Study in Foreign Economic Policy* (Stockholm: Almqvist and Wiksell, 1968), pp. 131-32; Stent, *Embargo to Ostpolitik,* pp. 93-127. The German pipe manufacturer Mannesmann lost $25 million, and Phoenix-Rheiner had to close a plant, as a result of the embargo. It is also possible that the United States opposed the pipeline for reasons of

economic self-interest. By the early 1960s, the USSR was increasing its oil exports—from 116,000 barrels a day in 1960 to 1 million barrels a day in 1965. Seventy percent of this was going to Western allies, particularly Italy, Japan, and West Germany. This meant a potential loss of business for American oil exporters, because the American barrel cost $2.50 and the Soviet barrel $1.75.

34. Pisar, *Coexistence and Commerce,* p. 63.

35. In 1964, the Senate Foreign Relations Committee conducted a survey of businessmen that showed a majority favored increased East-West trade. See Barnet, *The Giants,* p. 141.

36. *Department of State Bulletin,* December 2, 1964, p. 876.

37. Richard F. Kaufman, "Changing U.S. Attitudes Toward East-West Trade," in *NATO Colloquium on East-West Trade* (Brussels: NATO, 1983), pp. 3–4.

38. Ibid.

39. Office of Technology Assessment, *Technology . . . Trade,* p. 146.

40. U.S. Congress, *Public Law 91–184* (Washington, D.C.: U.S. Govt. Printing Office, 1969), 4(a)(1). The bill reaffirms the president's power to control exports. See also Beverly Crawford and Stefanie Lenway, "Decision Modes and International Regime Change: Western Collaboration on East-West Trade and the Trans-Siberian Pipeline Dispute" (paper delivered to the American Political Science Association, Chicago, 1983), p. 20.

41. Office of Technology Assessment, *Technology . . . Trade,* p. 115.

42. Stent, *Embargo to Ostpolitik,* p. 154.

43. Ibid., pp. 154–79; and Hanrieder and Auton, *Foreign Policies,* pp. 66–74.

44. Joan Edelman Spero, *The Politics of International Economic Relations* (New York: St. Martin's, 1981), pp. 310–11.

45. Gaddis, *Strategies,* pp. 274–309; LaFeber, . . . *Cold War,* pp. 258–80.

46. Barnet, *The Giants,* p. 141.

47. "Detente with the Soviet Union: The Reality of Competition and the Imperative of Cooperation," *Department of State Bulletin,* October 14, 1974, p. 508.

48. Henry Kissinger, *The White House Years* (Boston: Little, Brown, 1979), p. 1134.

49. John P. Hardt et al., *Western Investment in Communist Economies,* report to U.S. Senate, Committee on Foreign Relations (Washington, D.C.: U.S. Govt. Printing Office, 1974), pp. 61–83.

50. Ulam, *Dangerous Relations,* p. 93; Office of Technology Assessment, *Technology . . . Trade,* p. 147; Spero, *Politics . . . Relations,* p. 315.

51. *The Official Papers of Richard Milhous Nixon* (Washington, D.C.: U.S. Govt. Printing Office, 1972), p. 110.

52. Marshall Goldman, *Detente and Dollars* (New York: Basic Books, 1975), pp. 195, 212; Ulam, *Dangerous Relations,* p. 90.

53. Goldman, *Detente and Dollars,* p. 268; Spero, *Politics . . . Relations,* pp. 310–11.

54. Ulam, *Dangerous Relations,* pp. 122–23; Barnet, *The Giants,* p. 59.

55. U.S. Congress, 93rd Congress, 2nd sess. *Statutes at Large,* 88, pt. 2 (Washington, D.C.: U.S. Govt. Printing Office, 1974), pp. 2056–60.

56. Ulam, *Dangerous Relations,* pp. 122–23; U.S. Congress, *Statutes at Large,* pp. 2333–37.

57. U.S. Senate, Committee on Commerce, *Hearings: The American Role in East-West Trade* (Washington, D.C.: U.S. Govt. Printing Office, 1974), p. 2.

58. H. Heiss et al., "U.S.-Soviet Commercial Relations Since 1972," in U.S. Congress, Joint Economic Committee, *Soviet Economy in a Time of Change* (Washington, D.C.: U.S. Govt. Printing Office, 1979). See also Robert V. Roosa et al., *East-West Trade at a Crossroads* (New York: New York University Press, 1982), p. 18.

59. See Giscard's preface to Samuel Pisar, *Transactions Entre l'Est et l'Ouest* (Paris: Dunod, 1972). The French and British also were allegedly concerned that German-Soviet relations might get too friendly. See Kissinger, *The White House Years*, p. 422.

60. Ulam, *Dangerous Relations*, p. 193.

61. Ibid., p. 192.

62. Carter's pro-China policy also helped cool Soviet-American relations. In 1978, the president relaxed controls on exports to China, and in 1980 he granted the country MFN status.

63. A. J. Klinghoffer, "U.S. Foreign Policy and the Soviet Energy Predicament," *Orbis*, Fall 1981, p. 569.

64. Office of Technology Assessment, *Technology and East-West Trade: An Update* (Washington, D.C.: U.S. Govt. Printing Office, 1983), p. 104. Ironically, the vast majority of exceptions to the COCOM list have been granted to American companies.

65. Robert Paarlberg, "Lessons from a Grain Embargo," *Foreign Affairs*, Fall 1980, p. 152.

66. Robert J. Lieber, *The Oil Decade: Conflict and Cooperation in the West* (New York: Praeger, 1983), p. 139.

67. See, for example, statement by Senator Jake Garn, U.S. Congress, Senate, Committee on Banking, Housing, and Urban Affairs, *Hearings: Polish Debt* (Washington, D.C.: U.S. Govt. Printing Office, 1982), p. 1.

68. Crawford and Lenway, "Decision Modes," p. 23.

69. Kaufman, "Changing Attitudes," pp. 12-14.

70. Roosa et al, *East-West Trade*, p. 16.

71. "L'Ouest Condamné à Prêter à l'Est," *Le Matin* (Paris), March 30, 1982, p. 20; *Financial Times*, April 6, 1983.

72. *Financial Times*, November 12, 1982, and May 9, 1983.

73. Stephen Woolcock, *Western Policies on East-West Trade* (London: Royal Institute of International Affairs, 1982); Gary K. Bertsch, *East-West Strategic Trade, COCOM and the Atlantic Alliance* (Paris: Atlantic Institute, 1983), p. 26.

74. Bertsch, . . . *Strategic Trade*, pp. 26-27.

75. Hans Dietrich Genscher, "Toward an Overall Western Strategy for Peace, Freedom and Progress," *Foreign Affairs*, Fall 1982, p. 56; Agnelli, "East-West Trade," p. 1030; Office of Technology Assessment, *Technology . . . Trade*, p. 63.

CHAPTER 6

1. Alexander Haig, *Caveat: Realism, Reagan and Foreign Policy* (New York: Macmillan, 1984), p. 102.

2. Michael Ely, political counselor, U.S. embassy, interview, Paris, summer 1983; *The Economist*, August 15, 1981.

3. *Washington Post,* February 15, 1981.

4. Jonathan P. Stern, "Specters and Pipe Dreams," *Foreign Policy,* Fall 1982, p. 28; Haig, *Caveat,* p. 253; *New York Times,* August 6, 1981, p. D15.

5. Fabio Basagni, interview, Atlantic Institute, Paris, Summer 1983; Werber Nowak and Kurt Rippholz, "Surveying a Pipeline from Each End," *Wall Street Journal,* January 12, 1984, p. 25.

6. *Federal Regulations,* vol. 47, pp. 141–44 (1982); *New York Times,* December 30, 1981, p. 1.

7. Ibid.

8. Haig, *Caveat,* p. 254.

9. Jonathan B. Stein, *The Soviet Bloc, Energy and Western Security* (Lexington, MA: Lexington Books, 1983), pp. 63–64.

10. "Russian Gas Pipeline Loses GE as a Supplier; Only One French Company Could Fill the Gap," *Energy Daily* (New York), January 11, 1982, p. 1; Thane Gustafson, *Soviet Negotiating Strategy: The East-West Gas Pipeline Deal, 1980–1984* (Santa Monica, CA: Rand, 1985), p. 26.

11. Stein, *The Soviet Bloc,* p. 63.

12. Adam Ulam, *Dangerous Relations: The Soviet Union in World Politics, 1970–1982* (New York: Oxford University Press, 1983), p. 307; *New York Times,* December 15–29, 1981.

13. *International Herald Tribune,* June 15, 1982.

14. Ibid., February 8, 1982.

15. "Pressures on Poland," *Financial Times,* February 11, 1982.

16. Beverly Crawford and Stefanie Lenway, "Decision Modes and International Regime Change: Western Collaboration on East-West Trade and the Trans-Siberian Pipeline Dispute" (paper delivered to the American Political Science Association, Chicago, 1983), p. 23.

17. Testimony of Lionel Olmer, under secretary of commerce, U.S. Congress, House, Committee on Science and Technology, Subcommittee on Investigations and Oversight, *Hearings* (Washington, D.C.: U.S. Govt. Printing Office, 1982), p. 1.

18. Haig, *Caveat,* p. 304.

19. Interview, former State Department official who asked to remain anonymous.

20. James L. Buckley, under secretary of state, interview, fall 1985.

21. Crawford and Lenway, "Decision Modes," pp. 25–26.

22. Robert Hormats, assistant secretary of state for economic and business affairs, interview in New York City, June 1985.

23. Wilhelm Christians, interview, Düsseldorf, June 1985.

24. Buckley, interview.

25. Hormats, interview.

26. Haig, *Caveat,* pp. 304–05.

27. Ibid., p. 305.

28. Ibid., p. 306.

29. Hormats, interview; Haig, *Caveat,* p. 308.

30. Haig, *Caveat,* p. 309.

31. *Financial Times,* June 25, 1982.

32. Haig, *Caveat,* p. 309.

33. Ely, interview; *Financial Times,* June 25, 1982; *New York Times,* June 15, 1982.

34. Haig, *Caveat*, pp. 312–13.

35. Hormats, interview.

36. "Statement on Extension of U.S. Sanctions on the Export of Oil and Gas Equipment to the Soviet Union," *Weekly Compilation of Presidential Documents*, vol. 18, June 18, 1982; *Federal Regulations*, vol. 47, pp. 27, 250–51 (1982); *New York Times*, June 19, 1982, p. 1.

37. Ibid.

38. Haig, *Caveat*, pp. 252–53.

39. Richard Pipes, national security assistant for Soviet relations, interview, Cambridge, Massachusetts, Fall 1983.

40. *International Herald Tribune*, June 23 and July 2, 1982; *New York Times*, June 29, 1982; *Financial Times*, June 30, 1982.

41. Ely, interview.

42. Haig, *Caveat*, pp. 252–53.

43. Rudolf Augstein, publisher of *Der Spiegel*, interview, St. Tropez, France, Summer 1983.

44. *Financial Times*, June 30, 1982.

45. Stein, *The Soviet Bloc*, p. 68.

46. Ibid., p. 67.

47. *Le Monde*, July 20, 1982.

48. *Department of State Bulletin*, October 1982, pp. 40–41; *New York Times*, July 31, 1982.

49. Helmut Schmidt, chancellor, Federal Republic of Germany, interview, New York City, October 1985.

50. Samuel Pisar, *La Ressource Humaine* (Paris: Lattes, 1983), p. 167.

51. Office of Technology Assessment, *Technology and East-West Trade: An Update* (Washington, D.C.: U.S. Govt. Printing Office, 1983), p. 31; *New York Times*, November 13, 1982.

52. Central Intelligence Agency, "Outlook for the Siberian-to-Western Europe Natural Gas Pipeline," Sov 82-10120, Eur 82-10078 (1982).

53. U.S. Congress, Joint Economic Committee, *Soviet Pipeline Sanctions: The European Perspective* (Washington, D.C.: U.S. Govt. Printing Office, 1982), p. 2; *International Trade Reporter: U.S. Export Weekly*, July 20, 1982, p. 558; August 17, 1982, p. 699; and August 24, 1982, p. 748 (Washington, D.C.).

54. *Financial Times*, October 1, 1982.

55. Meyer Rashish, under secretary of state for economic affairs, interview, Washington, D.C., spring 1985.

56. Susan Haird, Department of Trade and Industry, interview, London, summer 1985.

57. *New York Times*, November 14, 1982. p. 1.

58. *Le Monde*, November 15, 1982, p. 1.

59. John F. Burns, "Progress for Soviet Pipeline," *New York Times*, July 29, 1983, p. D1.

60. U.S. Congress, House, Foreign Affairs Committee, *East-West Economic Issues, Sanctions Policy and the Formulation of International Economic Policy* (Washington, D.C.: U.S. Govt. Printing Office, 1984), p. 4.; John F. Burns, "Siberian-European Pipeline Project: Ambitious Goals, Doubtful Results," *New York Times*, January 6, 1984, p. 1.

61. Bill Paul, "Gas Glut May Trouble Soviet Pipeline," *Wall Street Journal,* January 19, 1984.

CHAPTER 7

1. U.S. Congress, *Public Law 96-72* (Washington, D.C.: U.S. Govt. Printing Office, 1979), 4(a).

2. *Wall Street Journal,* July 23, 1982.

3. Communauté Européen, *Gazoduc: Observations de la Communauté à l'Egard des Mesures Prises par le Gouvernement Américain* (Brussels: EEC, 1982), p. 3.

4. *Public Law 96-72*, 6(a-d).

5. Communauté Européen, *Gazoduc,* pp. 10–11.

6. A fascinating case involving extraterritorial reach involved Marc Rich and Co., a Swiss-based commodities firm. The Justice Department sought to subpoena Rich's records in its investigation of the company for alleged tax evasion. See "Court Decisions in the Marc Rich Case to Help U.S. Pursue Foreign Firms," *Wall Street Journal,* August 22, 1982, p. 17.

7. Donald DeKieffer, "Extraterritorial Application of US Export Controls—The Siberian Pipeline." American Society of International Law, *Proceedings of the 77th Annual Meeting* (Washington, D.C.: April 14-16, 1983), p. 243.

8. Stanley J. Marcuss, statement, U.S. Congress, Senate, Committee on Foreign Relations, Subcommittee on International Economic Policy, *Hearings: Soviet-European Gas Pipeline* (Washington, D.C.: U.S. Govt. Printing Office, 1982), p. 37. The pipeline embargo is even more difficult to justify under alternative governing "principles" of international law. For example, the "nationality principle" would allow American jurisdiction over U.S. corporate nationals. Yet most of the companies affected by the pipeline controls were incorporated in Europe. Moreover, there are no norms for determining the nationality of goods and technology under international law, and judicial decisions indicate that U.S. jurisdiction does not follow goods of U.S. origin once they enter the territory of another country. The "effects doctrine" could not have been applied, either, because exports from European countries to the Soviet Union could not have been said to have "direct, foreseeable and substantial" effects on U.S. trade, or even on the United States more broadly. Finally, the U.S. controls were a clear violation of the "territoriality principle," as the EEC's note of protest cited in the text demonstrates. See Patrick J. DeSouza, "The Soviet Gas Pipeline Incident: Extension of Collective Security Responsibilities to Peacetime Commercial Trade," *Yale Journal of International Law* 10 (Fall 1984): 111.

9. U.S. Congress, *The Export Administration Act Amendments of 1985,* Public Law 99-64 (Washington, D.C.: U.S. Govt. Printing Office, July 1985); Clyde H. Farnsworth, "New Focus on Soviet Trade Ties," *New York Times,* March 24, 1986, p. D1.

10. Douglas E. Rosenthal, statement, U.S. Congress, Senate, Committee on Foreign Relations, Subcommittee on International Economic Policy, *Hearings: Soviet-European Gas Pipeline,* pp. 45-46.

11. Office of Technology Assessment, *Technology and East-West Trade* (Washington, D.C.: U.S. Govt. Printing Office, 1979), pp. 120-23.

CHAPTER 8

1. U.S. Congress, House, Committee on Foreign Affairs, *East-West Economic Issues, Sanctions Policy and the Formulation of International Economic Policy* (Washington, D.C.: U.S. Govt. Printing Office, 1984), p. 46.

2. Office of Technology Assessment, *Technology and East-West Trade: An Update* (Washington, D.C.: U.S. Govt. Printing Office, 1983), p. 66.

3. International Energy Agency, "Communiqué from the Ministerial Meeting" (Paris: IEA, May 9, 1983); *Wall Street Journal,* May 9, 1983; *Financial Times,* March 30, 1983; Angela E. Stent, "Technology Transfers in East-West Trade: The Western Alliance Studies," *Foreign Policy and Defense Review,* January 1985, p. 49; U.S. Congress, *East-West Economic Issues,* p. 75.

4. *Wall Street Journal,* May 11, 1983.

5. Susan Haird, Department of Trade and Industry, interview, London, summer 1985.

6. *New York Times,* January 1, 1985, p. 24; January 27, 1985, p. E5.

7. W. Allen Wallis and Elliot Hurwitz, "A Collective Approach to East-West Economic Relations," unpublished paper (Washington, D.C.: Office of the Under Secretary of State for Economic Affairs, 1983); Office of Technology Assessment, *Technology . . . Update,* p. 67.

8. Clyde H. Farnsworth, "U.S. Approves the Sale of Oil Gear to Russians," *New York Times,* March 6, 1984, p. B1; Stent, "Technology Transfers," p. 50.

9. Mike Marshall, British representative to COCOM, interview, London, summer 1985.

10. Helmut Schmidt, former chancellor, Federal Republic of Germany, interview, New York City, October 1985.

11. *Department of State Bulletin,* September 1982, p. 38; *Journal of Commerce,* September 16, 1982, p. 1; Office of Technology Assessment, *Technology . . . Update,* p. 58.

12. Communauté Européen, *Gazoduc: Observations de la Communauté à l'Egard des Mesures Prises par le Gouvernement Américain* (Brussels: EEC, 1982), p. 2.

13. Fabio Basagni, interview, Atlantic Institute, Paris, Summer 1983; *Journal of Commerce,* December 29, 1982, p. 1.

14. Haird, interview.

15. Peter Rees, speech to the Royal Institute for International Affairs, Chatham House Conference on Extraterritoriality, London, October 21, 1982.

16. "Security Export Control," *British Business* (London), June 14, 1985, p. 2.

17. Michael Ely, former political counselor, U.S. embassy, interview, Paris, summer 1983.

18. Richard F. Kaufman, "Changing U.S. Attitudes Toward East-West Trade," in *NATO Colloquium on East-West Trade* (Brussels: NATO, 1983), p. 19.

19. Howard Lewis, testimony, U.S. Congress, Senate, Committee on Foreign Relations, *Workshop: The Premises of East-West Commercial Relations* (Washington, D.C.: U.S. Govt. Printing Office, 1982), p. 30.

CHAPTER 9

1. Herbert Stein, statement, U.S. Congress, Senate, Committee on Foreign Relations, Subcommittee on International Economic Policy, *Hearings: Economic Relations with the Soviet Union* (Washington, D.C.: U.S. Govt. Printing Office, 1982), p. 201.

2. Gunnar Adler-Karlsson, *Western Economic Warfare 1947–1967: A Case Study in Foreign Economic Policy* (Stockholm: Almqvist and Wiksell, 1968), pp. 22–82; Anthony C. Sutton, *Western Technology and Soviet Economic Development,* 3 vol. (Stanford, CA.: Stanford University Press, 1968–73).

3. Richard F. Kaufman, "Changing U.S. Attitudes Toward East-West Trade," *NATO Colloquium on East-West Trade* (Brussels: NATO, 1983), p. 20; Robert V. Roosa et al., *East-West Trade at a Crossroads* (New York: New York University Press, 1982), p. 20. The range goes from about 2 percent of Bulgarian GNP to 5 percent of Romanian GNP. Philip Hanson, *Trade and Technology in Soviet-Western Relations* (New York: Columbia University Press, 1981), p. 214.

4. CIA, memorandum from Maurice Ernst, National Intelligence Council, to Dennis Lamb, deputy assistant secretary for trade and commercial affairs, State Department, NIC #0021-83, January 3, 1983.

5. Boris Rumer and Stephen Sternheimer, "Soviet Economy: Going to Siberia," *Harvard Business Review,* January–February 1982.

6. Adam Ulam, *Dangerous Relations: The Soviet Union in World Politics, 1970–1982* (New York: Oxford University Press, 1983), p. 161.

7. Thane Gustafson, *Selling the Russians the Rope? Soviet Technology Policy and US Export Controls* (Santa Monica, CA.: Rand, 1981), pp. 67–68.

8. Office of Technology Assessment, *Technology and East-West Trade: An Update* (Washington, D.C.: U.S. Govt. Printing Office, 1983), p. 76.

9. Philip Hanson, "The Infusion of Imported Technology in the USSR," *East-West Technological Cooperation* (Brussels: NATO, 1976); Marshall Goldman, *USSR in Crisis: The Failure of an Economic System* (New York: W. W. Norton, 1983), p. 131.

10. Linda Malvern et al., *Techno-Bandits: How the Soviets Are Stealing America's High Technology Future* (Boston: Houghton Mifflin, 1984), pp. 54–56.

11. *Pravda,* February 24, 1981, as quoted in Goldman, *USSR in Crisis,* p. 133; *Pravda,* June 2, 1981, quoted ibid.; *Libération,* August 4, 1986, p. 6.

12. Goldman, *USSR in Crisis,* pp. 1–63.

13. Samuel Huntington, "Trade, Technology and Leverage," *Foreign Policy,* Fall 1978. Huntington is not a proponent of total trade denial.

14. *International Herald Tribune,* July 6, 1982.

15. Goldman, *USSR in Crisis,* pp. 29–63.

16. Henry Rowen, chairman, National Intelligence Council, statement, U.S. Congress, Joint Economic Committee, Subcommittee on International Trade, Finance and Security Economics, *Central Intelligence Agency Briefing on the Soviet Economy* (Washington, D.C.: U.S. Govt. Printing Office, 1982).

17. Office of Technology Assessment, *Technology . . . Update,* p. 52.

18. *New York Times,* August 4, 1985.

19. Giovanni Agnelli, "East-West Trade: A European View," *Foreign Affairs,* Summer 1980, p. 1028.

20. William Root, former director, Office of East-West Trade, Department of State, interview, Cambridge, Massachusetts, fall 1983. See also Beverly Crawford and Stefanie Lenway, "Decision Modes and International Regime Change: Western Collaboration on East-West Trade and the Trans-Siberian Pipeline Dispute" (paper delivered to the American Political Science Association, Chicago, 1983), p. 17.

21. Defense Science Board Task Force, *An Analysis of the Export Control of U.S. Technology—A DOD Perspective* (Washington, D.C.: U.S. Govt. Printing Office, 1976). This is known as the Bucy Report.

22. W. Clark McFadden II, Industry Coalition on Technology Transfer, statement, U.S. Congress, Senate, Committee on Governmental Affairs, Permanent Subcommittee on Investigations (Washington, D.C.: U.S. Govt. Printing Office, April 12, 1984), p. 8.

23. Samuel Pisar, *La Ressource Humaine* (Paris: Lattes, 1983), pp. 194–95. In an interview, Pisar described a problem of this nature that arose during the planning of the 1980 Olympic Games held in Moscow. Western planners realized that the computers they were sending to keep track of times and finishes could be used for military purposes.

24. Fabio Basagni, interview, Atlantic Institute, in Paris, summer 1983.

25. U.S. Congress, Senate, Committee on Governmental Affairs, Permanent Subcommittee on Investigations, *Transfer of United States High Technology to the Soviet Union and Soviet Bloc Nations* (Washington, D.C.: U.S. Govt. Printing Office, 1982), p. 577; Bill Keller, "U.S. Says Soviet Copies Some Arms," *New York Times,* September 19, 1985.

26. William T. Archey, acting assistant secretary of commerce, interview, Washington, D.C., fall 1984.

27. Department of Commerce, International Trade Administration, "Amendments to the Distribution License Procedure," Docket #40110, *Federal Register,* September 7, 1984.

28. Ibid.; see also National Association of Manufacturers, *Trade and Industry Newsletter,* June 15, 1984, p. 3.

29. Gov. Michael Dukakis, "News Release from the Office of the Governor" (Boston: Commonwealth of Massachusetts, April 3, 1984), p. 1.

30. Stein, statement, p. 199.

31. Howard Lewis, assistant vice-president for international trade, National Association of Manufacturers, interview, Washington, D.C., fall 1984.

32. McFadden, statement, p. 3.

33. Lewis, interview. Interviews with: Sue Eckert, Office of Congressman Don Bonker; Carole Rovner, Electronic Industries Association; Ron Fitzsimmons, Office of Congressman Les AuCoin; Chris Meinicke, Tektronix Corp.

34. Lewis, interview.

35. *Wall Street Journal,* July 24, 1984, p. 1.

36. Ibid.

37. Ibid.

38. Archey, interview.

39. Henry Kissinger, *The White House Years* (Boston: Little, Brown, 1979), p. 1272.

40. Richard Nixon, *Real Peace* (Boston: Little, Brown, 1983).

41. Robert Paarlberg, "Lessons from a Grain Embargo," *Foreign Affairs,* (Fall 1980), p. 145.

42. *Weekly Compilation of Presidential Documents,* January 28, 1980, p. 105; Office of Technology Assessment, *Technology . . . Update,* pp. 55–56.

43. See Gary C. Hufbauer and Jeffrey J. Schott, *Economic Sanctions Reconsidered: History and Current Policy* (Washington, D.C.: Institute for International Economics, 1985).

44. Prof. David Baldwin, Dartmouth College, interview, Cambridge, Massachusetts, fall 1983.

45. Samuel Pisar, *Coexistence and Commerce: Guidelines for Transactions Between East and West* (New York: McGraw-Hill, 1970), p. 73; Ulam, *Dangerous Relations,* pp. 94–95; Kaufman, "Changing U.S. Attitudes," pp. 20–24.

46. Agnelli, "East-West Trade," p. 1026.

47. Goldman, *USSR in Crisis,* p. 141; see also Ulam, *Dangerous Relations,* p. 16.

48. Serge Schmemann, "Concern for Secrecy in Soviet Inhibiting Use of Computers," *New York Times,* December 27, 1984, p. 1.

49. Richard J. Barnet, *The Giants: America and Russia* (New York: Simon and Schuster, 1977), p. 158.

50. Samuel Pisar, "A Human Dimension at Geneva," *International Herald Tribune,* November 19, 1985.

51. Heinrich Vogel, "Political-Security Implications of East-West Trade." To be published in *Challenges to the North Atlantic Alliance.*

52. Gary K. Bertsch, *East-West Strategic Trade, COCOM and the Atlantic Alliance* (Paris: Atlantic Institute, 1983), p. 28.

53. Samuel Pisar, interview, Paris, summer 1985.

CHAPTER 10

1. Helmut Schmidt, former chancellor, Federal Republic of Germany, interview, New York City, October 1985.

2. William A. Root, "Trade Controls That Work," *Foreign Policy,* Fall 1984, p. 77.

3. George Kennan, *Realities of American Foreign Policy* (Princeton: Princeton University Press, 1954), p. 87.

Bibliography

PUBLIC DOCUMENTS AND REPORTS

Amnesty International. *Prisoners of Conscience in the USSR: Their Treatment and Condition.* London: Amnesty International, 1981.

Central Intelligence Agency. Memorandum from Maurice Ernst, National Intelligence Council, to Dennis Lamb, deputy assistant secretary for trade and commercial affairs, Department of State. NIC #0021-83. January 3, 1983.

_____. "Outlook for the Siberian-to-Western Europe Natural Gas Pipeline." Sov 82-10120, Eur 82-10078. August 1982.

_____. "USSR: Development of the Gas Industry." Er 78-10393. 1978.

_____. *Prospects for Soviet Oil Production: A Supplemental Analysis.* Washington, D.C.: CIA, 1977.

Communauté Européen. *Gazoduc: Observations de la Communauté à l'Egard des Mesures Prises par le Gouvernement Américain.* Brussels: EEC, 1982.

Defense Science Board Task Force. *An Analysis of the Export Control of U.S. Technology–A DOD Perspective.* Washington, D.C.: U.S. Government Printing Office, 1976.

Department of Commerce, International Trade Administration. "Amendments to the Distribution License Procedure." *Federal Register,* September 7, 1984.

_____. "Amendment of Oil and Gas Controls to the USSR." *Federal Register,* June 24, 1982.

_____. Department of State. *Report to the U.S. Congress on Forced Labor in the USSR.* Washington, D.C.: Government Printing Office, 1983.

_____. *Department of State Bulletin.* December 2, 1964, p. 876.

"Detente with the Soviet Union: The Reality of Competition and the Imperative of Cooperation." *Department of State Bulletin,* October 14, 1974.

Dukakis, Michael. "News Release from the Office of the Governor." Boston: Commonwealth of Massachusetts, April 3, 1984.

"The European Community's Energy Strategy." *European File* no. 8 (1982).

The Export Administration Act Amendments of 1985. Public Law 99-64. Washington, D.C.: U.S. Government Printing Office, July 1985.

Federal Regulations, Vol. 47. Washington, D.C.: U.S. Government Printing Office, 1982.

"Forced Labour in the Soviet Union." *Background Briefs* (Foreign and Commonwealth Office), February 1983.

Goldman, Marshall. "Autarky or Integration: The USSR and the World Economy." In U.S. Congress, Joint Economic Committee. *Soviet Economy in a New Perspective.* Washington, D.C.: U.S. Government Printing Office, 1976.

Hanson, Philip. "The Infusion of Imported Technology in the USSR." In *East-West Technological Cooperation.* Brussels: NATO, 1976.

Hardt, John P., et al. *Western Investment in Communist Economies.* Report to the U.S. Senate, Committee on Foreign Relations. Washington, D.C.: U.S. Government Printing Office, 1974.

177

Heiss, H., et al. "U.S.-Soviet Commercial Relations Since 1972." In U.S. Congress, Joint Economic Committee. *Soviet Economy in a Time of Change.* Washington, D.C.: U.S. Government Printing Office, 1979.

International Energy Agency. "Communiqué from the Ministerial Meeting." Paris: IEA, May 9, 1983.

———. *Energy Policies and Programmes of IEA Countries–1982 Review.* Paris: OECD/IEA, 1983.

———. *OECD Quarterly Oil Statistics—1st Quarter, 1983.* Paris: OECD/IEA, 1983.

———. *Natural Gas: Prospects to 2000.* Paris: OECD/IEA, 1982.

International Monetary Fund, Bureau of Statistics. *International Financial Yearbook, 1983.* Washington, D.C.: IMF, 1984.

Kaufman, Richard F. "Changing U.S. Attitudes Toward East-West Trade." In *NATO Colloquium on East-West Trade.* Brussels: NATO, 1983.

Larkin, Richard X., and Edward M. Collins. Statement. U.S. Congress, Joint Economic Committee, Subcommittee on International Trade, Finance, and Security Economics. *Allocation of Resources in the Soviet Union and China—1981.* Washington, D.C.: U.S. Government Printing Office, 1981.

Masters, Charles D. "World Petroleum Resources—A Perspective," Open File Report 85–248. U.S. Geological Survey, Washington, D.C., 1985.

McFadden, W. Clark II. Statement. U.S. Congress, Senate, Committee on Governmental Affairs, Permanent Subcommittee on Investigations. Washington, D.C.: U.S. Government Printing Office, April 12, 1984.

National Association of Manufacturers. *Trade and Industry Newsletter*, vol. 2, no. 9 (June 15, 1984):2.

Nixon, Richard Milhous. *Official Papers.* Washington, D.C.: U.S. Government Printing Office, 1972.

Office of Technology Assessment. *Technology and East-West Trade: An Update.* Washington, D.C.: U.S. Government Printing Office, 1983.

———. *Technology and East-West Trade.* Washington, D.C.: U.S. Government Printing Office, 1979.

Organization for Economic Cooperation and Development. *Economic Surveys: Norway.* Paris: OECD, 1980.

———. *Energy Prospects to 1985.* Paris: OECD, 1974.

Peterson, Peter G. "U.S. Commercial Relations in a New Era." *U.S. Department of Commerce Report,* August 1972.

Radio Liberty Research. *Soviet Political Prisoners Today: Types of Punishment and Places of Detention.* Washington, D.C.: Radio Liberty Research, 1984.

Rees, Peter. Speech to the Royal Institute for International Affairs, Chatham House Conference on Extraterritoriality. London, October 21, 1982.

Rowen, Henry. Statement. U.S. Congress, Joint Economic Committee, Subcommittee on International Trade, Finance, and Security Economics. *Central Intelligence Agency Briefing on the Soviet Economy.* Washington, D.C.: U.S. Government Printing Office, 1982.

Royal Institute of International Affairs. *Policies and Strategies of Those Countries Which Export Natural Gas to Western Europe.* London: RIIA, 1983.

"Security Export Control." *British Business* (Department of Trade and Industry, London), June 14, 1982.

Sonatrach/Bechtel Corporation. *The Hydrocarbon Development Plan of Algeria: Financial Projections, 1975–2000.* San Francisco: Bechtel Corporation, 1978.

Stein, Herbert. Statement. U.S. Congress, Senate, Committee on Foreign Relations, Subcommittee on International Economic Policy. *Hearings: Economic Relations with the Soviet Union.* Washington, D.C.: U.S. Government Printing Office, 1982.

Stern, Jonathan P. "Western Forecasts of Soviet and Eastern European Energy over the Next Two Decades (1980–2000)." In U.S. Congress, Joint Economic Committee. *Energy in Soviet Policy.* Washington, D.C.: U.S. Government Printing Office, 1979–80.

Treml, Vladimir G. "Soviet Dependence on Foreign Trade." *NATO Colloquium on East-West Trade.* Brussels: NATO, 1983.

U.S. Congress. *Public Law 96-67.* Washington, D.C.: U.S. Government Printing Office, 1979.

———. *Statutes at Large,* Vol. 88, pt. 2. Washington, D.C.: U.S. Government Printing Office, 1974.

———. *Public Law 91-184.* Washington, D.C.: U.S. Government Printing Office, 1969.

U.S. Congress, House, Committee on Banking and Currency. *Hearings on H.R. 4293 to Extend the Export Control Act of 1949.* Washington, D.C.: U.S. Government Printing Office, 1969.

U.S. Congress, House, Committee on Foreign Affairs, Subcommittee on Europe and the Middle East and on International Economic Policy and Trade. *Hearings: East-West Economic Issues, Sanctions Policy and the Formulation of International Economic Policy.* Washington, D.C.: U.S. Government Printing Office, 1985.

U.S. Congress, House, Committee on Science and Technology, Subcommittee on Investigations and Oversight. *Hearings.* Washington, D.C.: U.S. Government Printing Office, February 5, 1982.

U.S. Congress, Joint Economic Committee. *Soviet Pipeline Sanctions: The European Perspective.* Washington, D.C.: U.S. Government Printing Office, 1982.

U.S. Congress, Senate, Committee on Banking, Housing, and Urban Affairs. *Hearings: Polish Debt.* Washington, D.C.: U.S. Government Printing Office, 1982.

———. *Hearings: Proposed Trans-Siberian Natural Gas Pipeline.* Washington, D.C.: U.S. Government Printing Office, 1981.

U.S. Congress, Senate, Committee on Commerce. *Hearings: The American Role in East-West Trade.* Washington, D.C.: U.S. Government Printing Office, 1974.

U.S. Congress, Senate, Committee on Foreign Relations. *Workshop: The Premises of East-West Commercial Relations.* Washington, D.C.: U.S. Government Printing Office, 1982.

U.S. Congress, Senate, Committee on Foreign Relations, Subcommittee on International Economic Policy. *Hearings: Soviet-European Gas Pipeline.* Washington, D.C.: U.S. Government Printing Office, 1982.

U.S. Congress, Senate, Committee on Governmental Affairs, Permanent Subcommittee on Investigations. *Transfer of U.S. Technology to the Soviet Union and Soviet Bloc Nations.* Washington, D.C.: U.S. Government Printing Office, 1982.

Weekly Compilation of Presidential Documents. Washington, D.C.: January 28, 1980 and June 18, 1982.

Zaleski, Eugene, and Helgegard Wienert. *Technology Transfer Between East and West.* Paris: OECD, 1980.

BOOKS

Adler-Karlsson, Gunnar. *Western Economic Warfare 1947-1967: A Case Study in Foreign Economic Policy.* Stockholm: Almqvist and Wiksell, 1968.

Aron, Raymond. *La République Impériale.* Paris: Fayard, 1973.

Barnet, Richard J. *The Giants: America and Russia.* New York: Simon and Schuster, 1977.

Below, Yuri, and Tamara Waldeyar. *Forced Labor on the Siberian Gas Pipeline.* Frankfurt: International Society for Human Rights, 1982.

Bertsch, Gary K. *East-West Strategic Trade, COCOM and the Atlantic Alliance.* Paris: Atlantic Institute, 1983.

Crovitz, Gordon. *Europe's Siberian Gas Pipeline: Economic Lessons and Strategic Implications.* London: Institute for European Defence and Strategic Studies, 1983.

DePorte, A. W. *Europe Between the Superpowers.* New Haven: Yale University Press, 1979.

Dienes, Leslie, and Lenway, Stephanie. "Decision Modes and International Regime Change: Western Collaboration on East-West Trade and the Trans-Siberian Pipeline Dispute." Paper delivered to the American Political Science Association, Chicago, 1983.

Elliot, I. F. *The Soviet Energy Balance.* New York: Doubleday, 1974.

Gaddis, John Lewis. *Strategies of Containment.* New York: Oxford University Press, 1982.

Gellman, Barton. *Contending with Kennan: Toward a Philosophy of American Power.* New York: Praeger, 1984.

Goldman, Marshall. *USSR in Crisis: The Failure of an Economic System.* New York: W. W. Norton, 1983.

———. *The Enigma of Soviet Petroleum: Half Empty or Half Full?* London: Allen and Unwin, 1980.

———. *Detente and Dollars.* New York: Basic Books, 1974.

Grosser, Alfred. *Les Occidentaux: Les Pays de l'Europe et les Etats-Unis Depuis la Guerre.* Paris: Fayard, 1978.

Gustafson, Thane. *Soviet Negotiating Strategy: The East-West Gas Pipeline Deal, 1980-1984.* Santa Monica, CA: Rand Corporation, 1985.

———. *Selling the Russians the Rope? Soviet Technology Policy and US Export Controls.* Santa Monica, CA: Rand Corporation, 1981.

Haig, Alexander. *Caveat: Realism, Reagan and Foreign Policy.* New York: Macmillan, 1984.

Hanrieder, Wolfgang, and Graeme P. Auton. *The Foreign Policies of West Germany, France and Britain.* Englewood Cliffs, NJ: Prentice-Hall, 1980.

Hanson, Philip. *Trade and Technology in Soviet-Western Relations.* New York: Columbia University Press, 1981.

Hewett, Ed A. *Energy, Economics and Foreign Policy in the Soviet Union.* Washington, D.C.: Brookings Institution, 1984.

Holzman, Frank. *International Trade Under Communism—Politics and Economics.* New York: Basic Books, 1976.

Hufbauer, Gary C., and Jeffrey J. Schott. *Economic Sanctions Reconsidered: History and Current Policy*. Washington, D.C.: Institute for International Economics, 1985.

Hutchings, Raymond. *The Soviet Budget*. New York: SUNY Press, 1983.

Kennan, George F. *Russia and the West Under Lenin and Stalin*. Boston: Little, Brown, 1961.

_____ . *Realities of American Foreign Policy*. Princeton: Princeton University Press, 1954.

Kissinger, Henry. *The Years of Upheaval*. Boston: Little, Brown, 1982.

_____ . *The White House Years*. Boston: Little, Brown, 1979.

Klinghoffer, Arthur J. *The Soviet Union and International Oil Politics*. New York: Basic Books, 1977.

LaFeber, Walter. *America, Russia and the Cold War*. New York: John Wiley and Sons, 1980.

LaVigne, Marie. *Les Relations Economiques Est-Ouest*. Paris: Presses Universitaires de France, 1979.

Lieber, Robert J. *The Oil Decade: Conflict and Cooperation in the West*. New York: Praeger, 1983.

Maull, Hanns W. *Natural Gas and Economic Security: Problems for the West*. Paris: Atlantic Institute, 1981.

_____ . *Europe and World Energy*. London: Butterworth, 1980.

Malvern, Linda, David Hebditch, and Nick Anning. *Techno-Bandits: How the Soviets Are Stealing America's High Technology Future*. Boston: Houghton Mifflin, 1984.

Nixon, Richard M. *Real Peace*. Boston: Little, Brown, 1983.

Nove, Alec. *East-West Trade: Problems, Prospects, Issues*. Washington, D.C.: Praeger, 1978.

Perez, Yves. *La Dissuasion par les Embargos*. Paris: Cahiers d'Etudes Stratégiques/ CIRPES, 1985.

Pipes, Richard. *Survival Is Not Enough*. New York: Simon and Schuster, 1984.

Pisar, Samuel. *La Ressource Humaine*. Paris: J. C. Lattes, 1983.

_____ . *Transactions Entre l'Est ou l'Ouest*. Paris: Dunod, 1972.

_____ . *Coexistence and Commerce: Guidelines for Transactions Between East and West*. New York: McGraw-Hill, 1970.

Putnam, Robert, and Nicholas Bayne. *Hanging Together: The Seven-Power Summits*. Cambridge, MA: Harvard University Press, 1984.

Revel, Jean-François. *Comment les Démocraties Finissent*. Paris: Grasset, 1983.

Roosa, Robert V., et al. *East-West Trade at a Crossroads*. New York: New York University Press, 1982.

Rosenblatt, S. *East-West Trade in Technology: A Purpose in Search of a Policy*. Washington, D.C.: Institute for International Economics, 1980.

Rosencrance, Richard. *The Rise of the Trading State*. New York: Basic Books, 1985.

Rustow, Dankert A. *Oil and Turmoil*. New York: Norton, 1982.

Shifrin, Avraham. *The First Guidebook to Prisons and Concentration Camps of the Soviet Union*. New York: Bantam Books, 1982.

Sokoloff, Georges, ed. *La Drôle de Crise: De Kaboul à Genève 1979–1985*. Paris: Fayard, 1986.

Spero, Joan Edelman. *The Politics of International Economic Relations*. New York: St. Martins, 1981.

Spiegel, Steven L. *The Other Arab-Israeli Conflict*. Chicago: University of Chicago Press, 1985.

Stein, Jonathan B. *The Soviet Bloc, Energy and Western Security.* Lexington, MA: Lexington Books, 1983.

Stent, Angela E. *Soviet Energy and Western Europe.* New York: Praeger, 1982.

―――― . *From Embargo to Ostpolitik: The Political Economy of West German-Soviet Relations, 1955-1980.* Cambridge: Cambridge University Press, 1981.

Stern, Jonathan P. *Soviet Natural Gas Development to 1990.* Lexington, MA: Lexington Books, 1980.

Sutton, Anthony C. *Western Technology and Soviet Economic Development.* 3 volumes. Stanford, CA: Stanford University Press, 1968-73.

Ulam, Adam. *Dangerous Relations: The Soviet Union in World Politics, 1970-1982.* New York: Oxford University Press, 1983.

―――― . *The Rivals.* New York: Oxford University Press, 1971.

Ullman, Richard H. *The Anglo-Soviet Accord: Anglo-Soviet Relations, 1917-1921.* Princeton: Princeton University Press, 1973.

Woolcock, Stephen. *Western Policies on East-West Trade.* London: Royal Institute of International Affairs, 1982.

ARTICLES

Agnelli, Giovanni. "East-West Trade: A European View." *Foreign Affairs,* Summer 1980.

Crawford, Beverly, and Stefanie Lenway. "Decision Modes and International Regime Change: Western Collaboration on East-West Trade and the Trans-Siberian Pipeline Dispute." Paper delivered to the American Political Science Association, Chicago, 1983.

DeKieffer, Donald. "Extraterritorial Application of US Export Controls―The Siberian Pipeline." *American Society of International Law: Proceedings of the 77th Annual Meeting* (Washington, D.C.: April 14-16, 1983), p. 243.

DeSouza, Patrick J. "The Soviet Gas Pipeline Incident: Extension of Collective Security Responsibilities to Peacetime Commercial Trade." *Yale Journal of International Law* 10 (Fall 1984).

Fabbri, Fabio. "The Balance of Economics and Politics in East-West Trade: The U.S. and Western Europe in East-West Economic Relations." Unpublished manuscript. Cambridge, MA: Harvard University, Center for International Affairs, 1983.

Fazzone, Patrick B. "Business Effects of the Extraterritorial Reach of the U.S. Export Control Laws." *Journal of International Law and Politics* 15 (Spring 1983).

Frost, Ellen L., and Angela Stent. "NATO's Troubles with East-West Trade." *International Security,* Summer 1983.

Genscher, Hans Dietrich. "Toward an Overall Western Strategy for Peace, Freedom and Progress." *Foreign Affairs,* Fall 1982.

Huntington, Samuel. "Trade, Technology and Leverage." *Foreign Policy,* Fall 1978.

Klinghoffer, Arthur J. "U.S. Foreign Policy and the Soviet Energy Predicament." *Orbis,* Fall 1981.

Lebahn, Axel. "The Yemal Gas Pipeline from the USSR to Western Europe in the East-West Conflict." *Aussen Politik* 34 (3rd quarter 1983).

Levy, Walter J. "Oil and the Decline of the West." *Foreign Affairs,* Summer 1980.

Meyer, Michael R. "Tapping the Sahara's Riches." *Newsweek,* July 18, 1983, p. 38.

Paarlberg, Robert. "Lessons from a Grain Embargo." *Foreign Affairs,* Fall 1980.

Root, William A. "Trade Controls That Work." *Foreign Policy,* Fall 1984.

Rumer, Boris, and Stephen Sternheimer. "Soviet Economy: Going to Siberia." *Harvard Business Review,* January–February 1982.

Stent, Angela E. "Technology Transfers in East-West Trade: The Western Alliance Studies." *Foreign Policy and Defense Review,* January 1985.

Stern, Jonathan P. "Specters and Pipe Dreams." *Foreign Policy,* Fall 1982.

Tucker, Robert W. "American Power and the Persian Gulf." *Commentary,* November 1980.

Vernon, Raymond. "The Fragile Foundations of East-West Trade." *Foreign Affairs,* Summer 1979.

Vogel, Heinrich. "Political-Security Implications of East-West Trade." To be published in *Challenges to the North Atlantic Alliance.*

Wallis, W. Allen, and Elliot Hurwitz. "A Collective Approach to East-West Economic Relations." Unpublished paper. Washington, D.C.: Office of the Under Secretary of State for Economic Affairs, 1983.

NEWSPAPERS AND PERIODICALS

Boston Globe
Economist
Le Figaro
Financial Times
International Herald Tribune
Libération
Le Matin
Le Monde
Newsweek
New York Times
Oil Daily
Petroleum Economist
Petroleum Intelligence Weekly
Pravda
Wall Street Journal
Washington Post

Glossary

NATO:	North Atlantic Treaty Organization
OECD:	Organization for Economic Cooperation and Development
IEA:	International Energy Agency
COCOM:	Consultative Group Coordinating Committee
EEC:	European Economic Community
CIA:	Central Intelligence Agency
NSC:	National Security Council
CMEA:	Council for Mutual Economic Assistance
OPEC:	Organization of Petroleum Exporting Countries
MFN:	Most Favored Nation
bcm:	billion cubic meters
BTU:	British Thermal Unit
mtoe:	million tons oil equivalent

Index

About the Author

Antony J. Blinken attends the Columbia University School of Law. He was a reporter for *The New Republic* magazine in Washington, D.C. Mr. Blinken is the author of numerous articles on foreign affairs and domestic politics. In 1986, he was selected as a fellow to the Salzburg Seminar on American Politics and the Foreign Policy Process.

Born in New York City, Mr. Blinken lived for ten years in Paris and received a French *baccalauréat* with high honors. He graduated magna cum laude from Harvard College, where he was executive editor of *The Harvard Crimson*.